BIOTECHNOLOGY, HUMAN NATURE, AND CHRISTIAN ETHICS

In public debates over biotechnology, theologians, philosophers, and political theorists have proposed that biotechnology could have significant implications for human nature. They argue that ethical evaluations of biotechnologies that might affect human nature must take these implications into account. In this book, Gerald McKenny examines these important yet controversial arguments, which have in turn been criticized by many moral philosophers and professional bioethicists. He argues that Christian ethics is, in principle, committed to some version of the claim that human nature has normative status in relation to biotechnology. Showing how both criticisms and defenses of this claim have often been facile, he identifies, develops, and critically evaluates three versions of the claim, and contributes a fourth, distinctively Christian version to the debate. Focusing on Christian ethics in conversation with secular ethics, McKenny's book is the first thorough analysis of a controversial contemporary issue.

GERALD MCKENNY is Walter Professor of Theology at the University of Notre Dame. He is the author of *To Relieve the Human Condition: Bioethics, Technology, and the Body* and *The Analogy of Grace: Karl Barth's Moral Theology*. His work in theological ethics and biomedical ethics is concerned with Christian ethics in a milieu that is shaped by modern culture, politics, and technology.

NEW STUDIES IN CHRISTIAN ETHICS

General Editor
Robin Gill

Editorial Board
Stephen R. L. Clark, Stanley Hauerwas, Robin W. Lovin

Christian ethics has increasingly assumed a central place within academic theology. At the same time, the growing power and ambiguity of modern science and the rising dissatisfaction within the social sciences about claims to value neutrality have prompted renewed interest in ethics within the secular academic world. There is, therefore, a need for studies in Christian ethics that, as well as being concerned with the relevance of Christian ethics to the present-day secular debate, are well informed about parallel discussions in recent philosophy, science, or social science. *New Studies in Christian Ethics* aims to provide books that do this at the highest intellectual level and demonstrate that Christian ethics can make a distinctive contribution to this debate – either in moral substance or in terms of underlying moral justifications.

TITLES PUBLISHED IN THE SERIES

(*continued after the Index*)

BIOTECHNOLOGY, HUMAN NATURE, AND CHRISTIAN ETHICS

GERALD McKENNY

University of Notre Dame

CAMBRIDGE
UNIVERSITY PRESS

CAMBRIDGE
UNIVERSITY PRESS

University Printing House, Cambridge CB2 8BS, United Kingdom

One Liberty Plaza, 20th Floor, New York, NY 10006, USA

477 Williamstown Road, Port Melbourne, VIC 3207, Australia

314-321, 3rd Floor, Plot 3, Splendor Forum, Jasola District Centre, New Delhi - 110025, India

79 Anson Road, #06-04/06, Singapore 079906

Cambridge University Press is part of the University of Cambridge.

It furthers the University's mission by disseminating knowledge in the pursuit of
education, learning and research at the highest international levels of excellence.

www.cambridge.org
Information on this title: www.cambridge.org/9781108435154
DOI: 10.1017/9781108385916

First published 2018
First paperback edition 2021

A catalogue record for this publication is available from the British Library

Library of Congress Cataloging in Publication data
Names: McKenny, Gerald, author.
Title: Biotechnology, human nature, and Christian ethics / Gerald McKenny,
University of Notre Dame.
Description: New York, NY: Cambridge University Press, [2018] |
Series: New studies in Christian ethics | Includes bibliographical references and index.
Identifiers: LCCN 2017046243 | ISBN 9781108422802 (hardback) |
ISBN 9781108435154 (paperback)
Subjects: LCSH: Biotechnology – Social aspects. | Ethics.
Classification: LCC TP248.23. M35 2018 | DDC 660.6–dc23
LC record available at https://lccn.loc.gov/2017046243

ISBN 978-1-108-42280-2 Hardback
ISBN 978-1-108-43515-4 Paperback

To the memory of Peter Andre (1984–2008)

Contents

General Editor's Preface

In 1992 Kieran Cronin's pioneering study *Rights and Christian Ethics* was the first monograph to be published in *New Studies in Christian Ethics*. It anticipated and helped shape Christian responses to a growing emphasis on human rights within health care and social ethics. It is very fitting that Gerald McKenny's fine contribution *Biotechnology, Human Nature, and Christian Ethics* now adds to this long-running series, seeking as it does to shape responses to the extraordinary emerging techniques that might one day make various human "enhancements" possible.

Like all other contributors to the series, Gerald McKenny has carefully observed the two central aims of *New Studies in Christian Ethics*, namely:

1. To promote monographs in Christian ethics that engage centrally with the present secular moral debate at the highest possible intellectual level.
2. To encourage contributors to demonstrate that Christian ethics can make a distinctive contribution to this debate – either in moral substance, or in terms of underlying moral justifications.

In *Moral Passion and Christian Ethics* (2017) I offered a critical analysis of Kieran Cronin's original contribution and used this as a standard for assessing all other books in the series – except, of course, the present one. I identified three slightly more refined phases in *Rights and Christian Ethics*. The first phase involved learning from a secular discipline (in this instance, philosophy and political theory) and encouraging other Christian ethicists to do likewise. The second phase involved challenging a purely secular understanding of the issue at hand (in this instance human rights) and deepening it with an understanding that is not entirely secular. And the third phase involved exploring a distinctively theological justification for moral choices and acts.

How does *Biotechnology, Human Nature, and Christian Ethics* measure up in terms of these three phases? I believe that it does so remarkably well. Gerald McKenny has listened attentively to a range of different voices within bioethics and taken proper account of the scientific and philosophical factors involved in the possibilities (or improbabilities) of different forms of human enhancement. He has identified three versions of the claim that normative status attaches to human nature in the context of biotechnology. This part of his task brings considerable clarity to secular and religious positions alike. He sees strengths and weaknesses in each and agrees that the standard bioethical principles of autonomy, safety, and fairness remain important. But, like many other Christian ethicists, he does not regard these principles as morally sufficient. So, he has suggested a fourth, more theological, version, grounded here specifically in Christology and eschatology, which seeks to challenge and deepen purely secular positions. Yet he has done this with admirable theological modesty. Here is a Christian apologist who is more than prepared to learn from secular discourse, albeit without being exclusively limited to its principles.

This is an impressive achievement, and I am delighted to welcome this percipient monograph to the series.

Robin Gill

Preface

In recent decades, bioethicists, policy makers, journalists, and others have been intrigued by biotechnological interventions that aim not at the prevention or treatment of diseases or injuries, but at desired changes in performances (athletic, cognitive, sexual), behaviors (moods, emotions, dispositions), and physical traits (longevity, height, physiognomy). These changes, which are commonly, if imprecisely, called enhancements, are brought about by a wide assortment of technologies. Notwithstanding their novelty and technical sophistication, some of these technologies, including pharmaceuticals (for example, anabolic steroids, Adderall, and Viagra) and surgical manipulations, are continuous with efforts at human self-transformation that reach back into the archaic human past. Other technologies, such as those that involve genetics, neuro-digital interfaces, and tissue-regenerative processes, are historically unprecedented. But in either case, these technologies are enabling us, at least to some extent, to control, select, bypass, replace, and alter human biological functions and traits to degrees and in ways that were never possible. It is unsurprising, then, that their development has been accompanied by recurring questions about human nature. In particular, many people ask whether the potential of biotechnology to implicate biological functions and traits in these ways requires us to take human nature into account in our ethical evaluations of biotechnology. Does human nature have some normative significance (that is, some meaning, value, purpose, or role) that ethical evaluations of biotechnology should consider? If so, then what is that significance and how should we take it into consideration? In short, what is the normative status of human nature, and how should it count in ethical evaluations of biotechnologies that implicate human nature?

These questions, and the answers that have been or may be given to them, are the subject matter of this book. At the end of the first chapter I offer a rationale for addressing them in a book-length inquiry in Christian ethics. But it is appropriate to say something here about the context in which they are to be understood. This context can be conceived narrowly to cover the past two or three decades, in which widely publicized developments – first in genetics and psychopharmacology, then in neuro-digital interfaces and regenerative medicine – have generated widespread and sometimes heated debates on the normative status of human nature. Nearly all the authors whose work I examine in this book belong to this period. However, the positions most of these authors take have antecedents in the work of authors (including Hannah Arendt, Hans Jonas, C. S. Lewis, Karl Rahner, and Pierre Teilhard de Chardin) who wrote in the decades immediately following World War II and whose concerns reflect developments in genetics and reproductive technology that occurred during those decades, along with the discrediting of eugenics. This broader context suggests that a common discourse on biotechnology and human nature runs through our era of biotechnology, in which optimism regarding scientific and technological progress is haunted by the memory of abuses of knowledge and technique. Taking a still broader context into account, the capability of biotechnology to intentionally intervene into human nature, and especially its capability of selecting, replacing, and altering biological functions and traits, can be understood as an event that is comparable in its significance to the modern "discovery" that social and political arrangements are not merely given but are subject to human construction. Of course, we have come to realize that social and political arrangements are far more recalcitrant to human construction than was initially suspected, and the same has proved true, at least up to now, in the case of human nature. It is also true that human beings have always sought and found ways to intervene into their nature despite its alleged givenness, just as they did with their social and political arrangements. These points should caution us not to exaggerate the novelty of the so-called biotechnological era. Nevertheless, intentional intervention into human biological nature is now a full-scale program, and this circumstance affects our attitudes and practices regarding our nature as well as our self-understanding as agents, even as the actual achievements of human biotechnology thus far remain modest and intermittent. It also poses the question whether the basis for the modern hope that human nature and the basic conditions of human life are fundamentally alterable and can be improved by human action, which has always oscillated between the realm of politics and that of science and technology (while usually taking both in), has come to rest, at least for now, in the latter realm (where Bacon and Descartes placed it) rather than the former. (For someone like me, whose official academic career began in 1989, it is at least symbolically significant that the first human gene therapy clinical trials as well as such an ambitious biotechnology project as the Human Genome Initiative were getting underway at about the same time as the last vestiges of a utopian political program were unraveling with the fall of the Berlin Wall and the collapse of the Soviet Union.)

If it is the case that hopes and ambitions regarding the human condition are increasingly directed to biotechnology, then the conversation about the normative status of human nature considering biotechnological enhancement belongs to a larger set of conversations, all of them characteristically modern, that focus on the proper stance toward human vulnerabilities and limitations in light of what is at least thought to be our increasing power to overcome them. Should humans commit themselves to those moral norms and virtues and social and political arrangements that enable them to live rightly and well with their characteristic vulnerabilities and limitations or to those that press them to overcome as many vulnerabilities and limitations as they can? Of course, these alternatives are not mutually exclusive; the question is what it would mean to do justice to both. My point, however, is that biotechnological enhancement is one of the most important contemporary sites in which the claims of both are contested and will be resolved, for better or worse.

Considerations drawn from this broader context take us beyond Christian ethics to the description, explanation, and criticism of modern culture. Those tasks lie beyond the scope of this book. But their relevance suggests that it is as important for Christian ethics to determine its relationship to contemporary biotechnology as it is to determine its relationship to modern political orders and movements. This book is a contribution to that task. It begins by clarifying how contemporary biotechnology implicates human nature and thus raises the question of its normative status, what it means to claim that normative status attaches to human nature, why that claim can be problematic, and why it is important for Christian ethics to try to make that claim in a way that avoids the problems that have always attended it (Chapter 1). It then goes on to formulate and critically examine three versions of that claim and to propose a fourth one (Chapters 2–5). Following a conclusion that presents the results of the inquiry, an appendix addresses the question of the implications of the normative status of human nature considering the distant but possible prospect of the transformation of human nature into something else.

Before turning to the main body of this inquiry, it is appropriate to say a word about its scope. It is not a general inquiry into the ethics of biotechnological enhancement or even an inquiry into the role of the normative status of human nature in resolving concrete ethical issues related to biotechnological enhancement. Still less is it an inquiry into the ethics of technology. Finally, it is not an inquiry into human nature. It is instead an inquiry into the normative status of human nature in light of the growing capabilities of contemporary biotechnological enhancement to implicate human nature. During this inquiry, much is said about the ethics of particular biotechnological enhancements, not a little is said about human nature, and something is said about the ethics of technology. But the emphasis of this book is on what it means to claim in the context of biotechnology that normative status attaches to human nature and how the normative status of human nature should count in the ethical evaluation of biotechnological interventions that implicate human nature. If the reader protests that this focus is too narrow, I would say two things in defense of it. First, there are already many fine studies in Christian ethics and related fields of the ethics of biotechnological enhancement, the ethics of technology, and human nature, but there is no thorough study in these fields of the normative status of human nature in relation to biotechnology. Second, if (as I am convinced) Christian ethics is in its own way as heavily invested in what biotechnology means for our biological nature as modern science, technology, and politics are in their way, then there is a need for a study in Christian ethics that takes it as seriously as it is taken in these other domains. It is my hope that this book will go some way toward meeting that need.

Acknowledgements

This book was prompted by two invitations, enabled by the material support of two institutions, assisted by various invitations to present it in part or in whole, improved by the contributions of many readers and interlocutors, and brought to the scholarly community by a visionary series editor.

The invitation that first prompted me to think about this topic came in 2004 from David Albertson and Cabell King, who as doctoral students led an interdisciplinary project on theology and nature at the University of Chicago Divinity School. At that time, prominent members of the US President's Council on Bioethics had recently published books defending the normative significance of human nature in the face of biotechnology. I did not intend to pursue the topic beyond the essay I contributed to the published volume and had moved on to other things when I received an invitation to deliver a keynote address to the 2012 meeting of the Society for the Study of Christian Ethics on a topic related to theological anthropology. By that time, more Christian ethicists were engaging the topic of biotechnology and human nature, while denials that human nature has normative status had gained in number and force in the field of bioethics. It seemed an appropriate time to embark on an investigation of the matter.

I could take up and complete the investigation thanks to the willingness of institutions to support it. A research leave during the 2011–2012 academic year enabled me to complete my initial research and writing. I am grateful to the University of Notre Dame, and in particular to John McGreevy, Dean of the College of Arts and Letters, and Matthew Ashley, Chair of the Department of Theology, for granting me the leave. During the 2016–2017 academic year, the Center of Theological Inquiry (CTI) funded a year of research and provided an ideal environment, thanks to which I could bring the book to completion. I am grateful to William Storrar, its director, for inviting me to become a member of the CTI; to the John Templeton Foundation and the National Aeronautics and Space Administration for funding CTI's Societal Implications of Astrobiology project, in which I was a participant; and to my fellow CTI members for conversations and colloquium sessions that generated new ideas and helped me refine arguments. I am also grateful to Dean McGreevy and Professor Ashley for granting me another year of leave to accept the CTI invitation.

This book has greatly benefited from various institutions that invited me to deliver lectures or lead colloquia or seminars on its topic. I am grateful to Baylor University (and especially to Paul Martens) for an invitation to deliver the Daniel B. McGee Endowed Lecture; to the Society for the Study of Christian Ethics (and especially to Susan Parsons and Robert Song) for an invitation to deliver a keynote address at its annual meeting; to the Princeton Religious Ethics Discussion Group (and especially to Frederick Simmons) for an opportunity to present parts of the manuscript for discussion; to the University of Aberdeen Department of Divinity and Religious Studies (and especially to Brian Brock and Michael Mawson) for devoting its Theological Ethics seminar to the book while in progress and inviting me to deliver a public lecture on it; to the Center for Bioethics, Health and Society of Wake Forest University (and especially to Ana Smith Iltis and Kevin Jung) for the invitation to deliver a public lecture; to Wycliffe College of the University of Toronto (and especially to Paul Allen of Concordia University) for an invitation to deliver a lecture at a symposium on theology and science; and to the Yale University Department of Religious Studies (and especially to Christine Hayes and Jennifer Herdt) for invitations to present overviews of the project to its Religious Studies Colloquium and its Religious Ethics Colloquium. I am also grateful to the many members of the audiences at these events whose questions and comments pressed me to clarify and rethink matters great and small.

Brian Brock, Kevin Jung, Michelle Marvin, Michael Mawson, and Stephen Pope read complete drafts of the manuscript with insight and care that far exceeded what any author can reasonably expect of colleagues and graduate students who render such a service. In addition, at different points along the way, I gained clarity or direction on different aspects of the topic thanks to conversations with Maria Antonaccio, Neil Arner, Jesse Couenhoven, Andrew Davison, Kathleen Eggleson, Andrew Forsyth, Stanley Hauerwas, Jennifer Herdt, Paul Martens, Douglas Ottati, Jean Porter, Paul Scherz, Frederick Simmons, Phillip Sloan, Kathryn Tanner, and Jonathan Tran. I am of course responsible for the shortcomings that remain, but the contributions of all these friends and colleagues have made it a much better book than it would have been.

The New Studies in Christian Ethics series has published much of the best scholarship in Christian ethics during the past quarter of a century and has been a major factor in the intellectual vitality of the field during this period. A debt of gratitude to the editor of the series, Robin Gill, is owed by everyone who works in this field, and especially by those whose work has appeared in this distinguished series. I am deeply grateful to Professor Gill for his unfailing support of the project.

Special thanks are due to Toy Bunnag and Erika McKenny for their encouragement, and to James Haring for preparing the index.

Finally, this book is dedicated to the memory of my nephew.

CHAPTER I

Biotechnology, Normative Status, Human Nature

If biotechnology aims not only at the prevention and treatment of diseases and injuries but also at desired changes in our performances, behaviors, and physical characteristics, and if these changes involve our nature as human beings, then the ethics of biotechnology cannot avoid asking about the normative status of human nature. Does our nature have some normative significance that we must take into account when we consider biotechnological interventions that involve our nature? This question is not an esoteric one, of interest only to bioengineers and professors of ethics. People who follow developments in biotechnology or hear of them intermittently as they are reported in the media may find themselves wondering what all this means for us as human beings and whether whatever it does to us as human beings matters. However, it is a difficult question, and the twofold purpose of this chapter is to get clear about how biotechnology involves our nature, what it means to speak of the normative status of our nature, and why some people think it is problematic to attach normative status to human nature on the one hand and what is at stake for Christian ethics in this whole matter on the other hand.

I begin this chapter by identifying the various ways in which biotechnology implicates human nature. I then say why I think that the normative status of human nature is a matter of importance for Christian ethics, and I go on from there to introduce four views of the normative status of human nature that the following chapters consider in detail. Next, I try to express more precisely what it means to say that normative status attaches to human nature in the context of biotechnology. I then turn to the major criticisms of attempts to attach normative status to human nature. Finally, I set out what I think is ultimately at stake for Christian ethics in the normative status of human nature in the context of biotechnology.

How Biotechnology Implicates Human Nature

An inquiry into the normative status of human nature in the context of biotechnology should begin by clarifying how biotechnological interventions implicate human nature.[1] The following classification enumerates five ways in which current and prospective biotechnological interventions do this. Although it is not exhaustive, the list includes the ways that are most relevant to claims that normative status attaches to human nature:

1. Many pharmacological interventions, including anabolic steroids, selective serotonin reuptake inhibiters (SSRIs), and concentration enhancers (for example, Adderall or modafinil), as well as some neural-digital interface devices (for example, those that link brains with computers), do not change existing biological functions and traits (at least not profoundly) but rather exercise temporary *control* over them to achieve a desired state or performance.

2. In the realm of reproduction and genetics, techniques of gamete selection (as may occur in artificial insemination and in vitro fertilization) and embryo selection (as occurs in preimplantation genetic diagnosis) make it possible to *select* desired genetic characteristics by choosing for fertilization or implantation those gametes or embryos that possess the desired characteristics.

3. Other current or prospective reproductive technologies such as in vitro fertilization and reproductive cloning *bypass* biological functions or traits (in this case, the sexual function that is exercised in the conception of offspring) without acting on them.

4. In at least some cases, neural implants and mechanical prosthetics *replace* biological functions, doing (presumably in a superior or more desirable way) digitally or mechanically what was previously done biologically.

5. Finally, there is a wide variety of prospective interventions that aim to bring about heightened cognitive abilities; new or expanded perceptual capacities; a wider, narrower, or more intense emotional range; greatly increased physical strength or agility; or a vastly extended life span, and that thus *alter* biological functions or traits by bringing about permanent quantitative or qualitative changes to them.

---

[1] The phrase "implicates human nature" sounds vague, but it covers all the items in the following classification, which includes not only biotechnologies that act directly on biological functions and traits but also those that substitute for biological functions.

Two additional classifications pertain to this last category. First, there are changes *to* human nature (which remains human, albeit in an altered form) and changes *of* human nature into something else (as in "transhumanist" or "posthuman" scenarios in which human beings become something other than human). Second, there are changes that involve individuals only (so that *their* nature will have changed in one of the prior two ways) and changes that have population-level effects (so that human nature itself will have changed in one of the two prior ways). In some contexts involving the alteration of human nature, one or both classifications are significant; in other contexts, neither is significant. But they are especially important to keep in mind (as they too often are not) when dramatic alterations of human nature are under consideration.

In sum, biotechnology may control, select, bypass, replace, or alter aspects of human nature (or, in the last case, human nature itself). In any of these ways of implicating human nature, it may put the normative status of human nature at stake, but in the four positions I consider in the following chapters that status is put at stake mostly by the selection, replacement, or alteration of human functions or traits.

Human Biological Nature and Christian Ethics

Why should any of these ways of implicating human nature matter for Christian ethics? Why shouldn't Christian ethicists simply concern themselves with the ends of these interventions (that is, whether they promote genuine goods) and the means to those ends (that is, whether they violate any moral requirements in the pursuit of those goods)? Why take their implications for human nature as a theme? The premise of this book is that the implications of biotechnology for human nature cannot be a matter of indifference for Christian ethicists because human nature is not a matter of indifference to Christian ethics. To be sure, the significance of human biological nature, which is what is directly at stake in biotechnology, is for Christian ethics a qualified significance. Human biological nature is not the whole of human creaturely nature, which includes characteristics that are not reducible to biology. Whether the notion of a soul is necessary to render these characteristics fully intelligible or only a notion of emergent or supervenient properties, Christian theology understands human nature as a complex reality that cannot be reduced to biology. Moreover, for most Christian theologies, human nature as we experience it is not simply identifiable with the nature God created; like the rest of creation, it suffers the effects of the fall. These two points qualify the significance for Christian ethics of what biotechnology might do to implicate human biological nature. The biological nature on which biotechnology acts is not the only aspect of human nature to be considered when evaluating those acts, and the effects of the fall must also be considered in any such evaluation. Nevertheless, human biological nature is an important component of the human nature that, along with the rest of creation, God pronounced good and destined for eschatological fulfillment, and notwithstanding the effects of the fall, it is still part of the human nature that God created. For these two reasons, it matters to Christian ethics.

Of course, few Christian ethicists would deny this point. But they seldom give it its due. When they turn their attention to the creation narratives of Genesis 1 and 2, for example, Christian ethicists, like Christian theologians more generally, typically focus on those aspects of human nature and vocation that distinguish humans from other biological creatures. The divine grant of dominion over creation (Gen. 1:26–28), the vocation of tilling and keeping the Garden (Gen. 2:15), and the recognition of the other as one's fellow human (Gen. 2:23) all play prominent roles in Christian ethical reflection on the relationships of humans to one another and to the natural world. Less attention is paid to the overall picture of humanity that emerges from the striking depictions in these creation narratives of humans as biological creatures – depictions that are even more striking when they accompany a depiction of humans as godlike. Like other creatures of the earth, humans are characterized by sexual reproduction (Gen. 1:28); like other creatures of the sixth day, they are dependent for their survival on metabolic exchanges with other biological life (Gen. 1:29); and they are mortal beings who return to the earth from which they were formed (Gen. 2:7). Sexual reproduction, metabolism, mortality – whatever else they also are, humans are biological creatures.[2]

As biological creatures, humans are also destined (along with the rest of creation) for eschatological fulfillment. Scripture is ambiguous regarding the eschatological status of our biological nature. Is it transcended, as might be asserted of human sexual nature in Jesus' statement that "in the resurrection they neither marry nor are given in marriage, but are like angels in heaven" (Mt. 22:30)? Is it rather a matter of continuity, as we might infer from the eating and drinking of the resurrected Christ (Lk. 24:41–43)? Or

---

[2] My point is not that Christian ethicists ignore or fail to appreciate these features of human nature. It is hardly the case that sexual reproduction and mortality have been neglected, and now perhaps eating is beginning to receive due attention. My point is simply that features such as these have not been incorporated into a conception of human biological nature as a distinctive matter of normative significance.

is it in some way transformed, as the Apostle Paul seems to say of mortality in I Corinthians 15? It is not surprising that Christians have held a variety of positions on the eschatological status of human biological nature, but it is also not surprising that most Christians have resisted the idea that our biological nature is excluded from our eschatological fulfillment.[3]

Of course, a few simple citations of Scripture taken out of context do not begin to do justice to the extraordinary complexity of these matters, which will become apparent enough in the chapters that follow this one. Nor do they supply us with an explicit conception of human biological nature. They do, however, suggest that what we understand as human biological nature is a component of the human nature that God created good and destined for eschatological fulfillment, and that it is therefore not a matter of indifference to Christian ethics. And this suggestion – or so I will argue – is one that Christian ethics is not free to ignore in the context of biotechnology. To state the point positively and in the technical terms that the rest of this book will employ, for Christian ethics *normative status attaches to human biological nature*, which is to say that human biological nature counts in the ethical evaluation of human actions that implicate it. To say only that human biological nature *counts* in the evaluation of actions that implicate it is a deliberately weak claim that says nothing about how much it counts, but for now I only want to assert two claims that this book will establish, namely, that Christian ethics is not indifferent to actions that implicate human biological nature, and that to the extent that biotechnological actions are among these actions, Christian ethics cannot avoid coming to terms with the normative status of human biological nature in the context of biotechnology. To once again put the issue in positive terms, a necessary task of Christian ethics in relation to biotechnology is to determine what the normative status of human biological nature is and how it counts in the evaluation of biotechnological interventions that implicate human biological nature.

Christian ethicists who take on this task quickly find themselves amid a lively, ongoing debate in which Christian ethics participates but that also extends far beyond this field.[4] However, if they seek to formulate and

---

[3] A treatment of the nature of the postresurrected human being as it has been conceived in patristic and medieval Christian theology is, of course, well beyond the scope of this inquiry.

[4] The question of the normative status of human nature considering biotechnology has been taken up in the following books, all of them published since 2000: Nicholas Agar, *Humanity's End: Why We Should Reject Radical Enhancement* (Cambridge, MA: MIT Press, 2010); David Albertson and Cabell King, eds., *Without Nature? A New Condition for Theology* (New York: Fordham University Press, 2010); Harold W. Baillie and Timothy K. Casey, eds., *Is Human Nature Obsolete? Genetics, Bioengineering, and the Future of the Human Condition* (Cambridge, MA: MIT Press, 2005); Allen Buchanan, *Beyond Humanity? The Ethics of Biomedical Enhancement* (Oxford: Oxford University

defend the claim that normative status attaches to human nature in the context of biotechnology, they should not expect guidance or even sympathy from the academic field of bioethics. Most bioethicists today flatly deny that normative status attaches to human nature. Some of them argue that we are in principle at liberty to do as we wish in biotechnological engagements with our nature, while others argue that we are under a broad obligation to make use of biotechnology to promote the well-being of humans regardless of what becomes of our nature in the process. In either case, they hold that actions that implicate human biological nature are constrained only by generally applicable moral principles that see to it that the choices made by individuals are self-determining and not coerced (autonomy), that interventions meet an acceptable ratio of risks to benefits (safety), and that access to the relevant technologies is fair (fairness). Considerations having to do with human nature do not feature in their ethical deliberations.[5]

Press, 2011); Mark Coeckelbergh, *Human Being @ Risk: Enhancement, Technology, and the Evaluation of Vulnerability Transformations* (Dordrecht: Springer, 2013); Ronald Cole-Turner, ed., *Transhumanism and Transcendence: Christian Hope in an Age of Technological Enhancement* (Washington, DC: Georgetown University Press, 2011); Celia Deane-Drummond and Peter Manley Scott, eds., *Future Perfect? God, Medicine, and Human Identity* (Edinburgh: T&T Clark, 2006); Francis Fukuyama, *Our Posthuman Future: Consequences of the Biotechnology Revolution* (New York: Farrar, Straus and Giroux, 2002); Jürgen Habermas, *The Future of Human Nature* (Cambridge: Polity Press, 2003); Philip Hefner, *Technology and Human Becoming* (Minneapolis: Fortress Press, 2003); Malcolm Jeeves, ed., *Rethinking Human Nature: A Multidisciplinary Approach* (Grand Rapids, MI: Eerdmans, 2011); Paul Jersild, *The Nature of Our Humanity: A Christian Response to Evolution and Biotechnology* (Minneapolis: Fortress Press, 2009); Gregory Kaebnick, ed., *The Ideal of Nature: Debates about Biotechnology and the Environment* (Baltimore: The Johns Hopkins University Press, 2011); James C. Peterson, *Changing Human Nature: Ecology, Ethics, Genes, and God* (Grand Rapids, MI: Eerdmans, 2010); Michael J. Sandel, *The Case against Perfection: Ethics in the Age of Genetic Engineering* (Cambridge, MA: Harvard University Press, 2007); Tamar Sharon, *Human Nature in an Age of Biotechnology: The Case for Mediated Posthumanism* (Dordrecht: Springer, 2014); Allen Verhey, *Nature and Altering It* (Grand Rapids, MI: Eerdmans, 2010); Brent Waters, *From Human to Posthuman: Christian Theology and Technology in a Postmodern World* (Aldershot: Ashgate, 2006).

[5] I say more in the following text about the stance of mainstream bioethics toward the claim that normative status attaches to human nature. Most criticisms of the claim are found in brief dismissals of it or critical reviews of books by authors who defend it, but a thorough critical treatment of it is found in Buchanan, *Beyond Humanity?* A progenitor of contemporary critics of the claim is John Stuart Mill, who famously identified two concepts of nature, one of which includes "the sum of all phenomena, together with the causes which produce them," and the other of which opposes natural to artificial and thus includes "whatever takes place without the voluntary agency of man." As Mill points out, every human action accords with nature in the first sense and every human action violates nature in the second sense. Mill concludes that nature is unsuited to be a normative guide: According to the first sense, whatever we do, whether good or bad, right or wrong, is in accordance with nature, while in the second sense, whatever we do violates nature. See Mill, "On Nature," in *Essential Works of John Stuart Mill*, edited by Max Lerner (New York: Bantam, 1961), pp. 367–401. The problem with Mill's analysis is that it excludes the concept of nature as an order created by God, which in one form or another is central to the understanding of nature in many theistic traditions.

Contrasting with this mainstream bioethical stance, however, is a strong and persistent countercurrent represented by certain Christian, Jewish, and secular thinkers who in most cases are not professional bioethicists. Most of these thinkers agree with mainstream bioethicists that biotechnological interventions are subject to principles of autonomy, safety, and fairness. But they also hold that normative status attaches to human biological nature and that this status is relevant to the ethical evaluation of human biotechnology. They defend this status either by appealing to some meaning, value, or purpose of human biological nature or by presenting human biological nature as a condition of goods, rights, identity, or agency that biotechnology may imperil or promote. (The claim that biotechnology may promote such values rather than imperiling them indicates what Chapters 3 and 4 will demonstrate, namely, that some versions of the claim that normative status attaches to human nature are favorable to biotechnological enhancement.) Christian ethicists have their own grounds for ascribing normative status to human biological nature, as I have already noted and as Chapters 2 through 5 of this book will elaborate. Because this book is an inquiry in the field of Christian ethics, it is principally a critical elucidation and development of these grounds and their implications. However, claims regarding the normative status of human nature by Christian theologians and ethicists cannot be adequately understood apart from similar claims that have been made by Jewish and secular thinkers, so that any examination of how the former have carried out their task would be incomplete without attention to the body of literature produced by these thinkers, who have their own reasons for attaching normative status to human biological nature. Moreover, inasmuch as this book not only examines what Christian theologians and ethicists *have* said about the normative status of human nature but also makes a proposal about what they *should* say, it stands ready to receive instruction and correction from others who have attempted from their own perspectives to say what the normative status of human nature is and what relevance it has to biotechnology, and even to learn how to avoid problematic formulations of the normative status of human nature from the criticisms of bioethicists who reject that status. The point is that while there are reasons internal to Christian ethics for carrying out an inquiry into the normative status of human nature, there are also reasons for conducting that inquiry as a conversation with interlocutors from Jewish and secular bioethics, and that is what this book does.

I have drawn a distinction between a bioethical mainstream that denies that normative status attaches to human nature and a countercurrent that affirms it, but I want to warn against the temptation to divide the mainstream and countercurrent along familiar lines. The mainstream position, we might assume, is secular and favorable toward biotechnology, while the countercurrent is religious and opposed to biotechnology. However, that assumption is mistaken on both counts. First, regarding the secular-religious axis, the most prominent figures who represent the countercurrent are in fact secular thinkers who explicitly disavow religious grounds for their positions, while most Christian and Jewish bioethicists who write about human biotechnology fall clearly in the mainstream. Second, regarding the favorable-opposed axis, some important countercurrent figures hold that the particular normative significance human nature has makes biotechnological determination of human nature at least permissible and possibly obligatory, while it is at least in principle possible to argue against the determination or alteration of human functions and traits from within the mainstream by insisting on strict criteria for autonomy, safety, and fairness. In short, the mainstream includes religious voices and voices that oppose biotechnological determination of human nature, while the countercurrent includes secular voices and voices that strongly support biotechnological determination of human nature. These complex alignments of religious and secular voices and proponents and opponents of human biotechnology are reflected in the chapters that follow, in which Christians who claim that human biological nature has normative significance will be joined by some prominent secular thinkers (who are opposed or ignored by most Christian bioethicists in the mainstream), while they will disagree over whether the normative significance of human nature allows or disallows biotechnological determination or alteration of human nature (a disagreement in which both sides find support from mainstream bioethics). In sum, what divides the mainstream and the countercurrent is whether normative status attaches to human nature. Differences over religious and secular and acceptance of or opposition to biotechnology cut across this division.

Normative Status of Human Nature: Four Views

In the four chapters that follow (that is, Chapters 2–5), I formulate and critically examine what I take to be four distinct versions of the claim that normative status attaches to human nature in the context of biotechnology. My reconstructions, analyses, and critiques of these positions are my attempt to fulfill what I have just said is a necessary task of Christian ethics in relation to biotechnology, namely, to determine what the normative status of human nature is and how it counts in the evaluation of biotechnological interventions that implicate human nature.

The first version (NS1) appeals to people who want biotechnology to leave human nature as it is. It claims that normative status attaches to human nature as that which exists apart from intentional human action. On this view, acting to heal or restore human nature leaves it as it is, but acting to select or alter human functions or traits does not; it therefore violates the normative status of human nature. The typical rationale for this position is not, as one might suppose, the conviction that human biological nature is sacred and therefore must not be touched. It is rather the conviction that only if one has a biological nature that has not been chosen or altered by others can one be treated or recognized as the ontological or moral equal of others, be able to speak and act in one's own person, or be loved simply for who one is. For some subscribers to NS1, this person-centered conviction is accompanied by the additional conviction that intentional intervention into human biological nature involves a problematic attitude or stance in which one wrongly objectifies nature or treats it as mere raw material for human ambitions and desires. This additional conviction may seem identical to the conviction that nature is sacred, but the attitude or stance may be proscribed because it is an affront to God, not nature, or because it expresses a problematic desire for mastery; in either case, it is a nature-regarding but not a nature-centered conviction. Early versions of one or both claims were put forth by Hannah Arendt, Hans Jonas, and C. S. Lewis; more recent proponents (who are the focus of Chapter 2) are Jürgen Habermas, Oliver O'Donovan, and Michael Sandel.

The second version (NS2) focuses on human nature as the condition for human goods and rights. Subscribers to NS2 claim that normative status attaches to human nature as the ground of human goods or rights, and some of them worry that these goods or rights may be imperiled by the alteration of biological functions and traits. If, as they believe, human goods and rights are not simply constructed by society but are grounded in human nature, what becomes of them when biotechnology makes it possible to bypass or alter biological functions and traits? Will persons with different functions and traits have different rights? Will interventions into our nature put our overall well-being at risk or forfeit desires and attachments to others that depend on our vulnerabilities and limitations in favor of the enticing but ultimately superficial desires and attachments that accompany the overcoming of vulnerabilities and limitations? Francis Fukuyama, Leon Kass, and Martha Nussbaum all argue that human rights or the worthiest human goods depend in some way on our nature as it now is, while others argue that biotechnology is not necessarily a threat to nature-dependent goods and may even help us to realize them more fully. In both cases, it is the dependence of goods or rights on human nature that marks NS2.

The third version (NS3) appeals to many people who support biotechnological enhancement. It claims that normative status attaches to human nature as indeterminate, open-ended, or malleable. Some theologians who subscribe to NS3 point to these characteristics of human nature in arguing that humans are authorized to employ biotechnology to fulfill their God-given vocation to bring human nature to its completion or perfection, or at least to its next stage. Pierre Teilhard de Chardin and Karl Rahner were early proponents of this claim that has recently been championed by Philip Hefner, James Peterson, and Laurie Zoloth. Other subscribers to NS3, including Donna Haraway, stress the emancipatory possibilities that are opened by this understanding of human nature while also showing how the indeterminacy, open-endedness, and malleability of human nature undermine the support which notions of human nature as fixed and bounded have lent to racism, sexism, and other problematic "isms."

These three versions of the claim that normative status attaches to human nature are widely represented in the literature on the ethics of human enhancement. However, they have not been categorized in an intelligible way, and in some cases they have not been presented in their most plausible formulations. As a result, the claim that normative status attaches to human nature is not well understood, and both criticisms and defenses of this claim are often facile. Chapters 2 through 4 offer a systematic presentation and a thorough critical examination of these claims in the hope of bringing clarity to them and assessing their viability. Chapter 5 introduces into the debate a new version (NS4), which claims that normative status attaches to human nature as that which suits or equips humans for a certain form of life with God and with other human beings. In this form of life with God and other humans, I will argue, humans image God, and they have been given their particular biological characteristics by God because these are precisely the characteristics that suit them for the form of life with God and with other humans for which God has created them. Normative status thus attaches to human nature as a condition for this form of life with God and with other humans. Although it is my construction, this claim is strongly influenced by the views of the human being as creature propounded by Karl Barth and Kathryn Tanner.

Because this book is an inquiry in Christian ethics, it is appropriate to point out in advance the theological grounds of each of these claims regarding the normative status of human nature as they appear in those authors who write from a Christian perspective. In its conviction that human nature is to be kept off-limits to intentional human intervention, NS1 is consistent with a broadly Augustinian account of creation as a finished work whose eschatological transformation is not accomplished or approximated in historical time or through creaturely agency, including that of biotechnology. Biotechnology may help to restore aspects of fallen creation (as it does when it prevents or treats diseases and injuries), but it does not continue the work of creation or approximate the eschatological transformation of creation (as some might suppose it could do by selecting, replacing, or altering human functions or traits). In its grounding of goods and rights in human nature, NS2 is consistent with Aristotelian-Thomist accounts for which the goodness and intelligibility of creation is found in the ordering of the goods of different kinds of things to the kinds of thing they are. Human goods fulfill human beings as the kind of being they are, while human rights protect the pursuit of these goods, and the question posed to biotechnology is whether it promotes or imperils these natural goods and rights. In its emphasis on the susceptibility of human nature to human action aimed at its perfection (that is, on the indeterminacy, open-endedness, and malleability of human nature that allow for intervention into human functions and traits), NS3 is consistent with a broadly (and perhaps exaggerated) Irenaean account of creation as an unfinished work to be completed through a temporal process in which creaturely action plays a crucial role. From this perspective, biotechnology may be thought of as part of the human vocation to continue the work of creation and also approximate (or even accomplish) its eschatological transformation. Finally, in its conviction that human nature equips humans for life with God, NS4 draws heavily on certain patristic and Barthian claims about the relation of human nature to the *imago dei*. On this account, human nature offers the conditions in which the distinctively human vocation to conform to Christ as the image of God is to be realized, and the question posed to biotechnology is whether it instantiates or conflicts with the realization of that vocation.

These theological grounds are more or less explicitly articulated by the Christian subscribers to NS1, NS2, NS3, and NS4. Part of my task in Chapters 2 through 5 is to make them explicit where they are only implicit. But even where they are not fully articulated, they provide evidence that these four versions of the claim that normative status attaches to human nature are not mere ad hoc responses to biotechnology but are deeply rooted in Christian tradition. I will argue in the conclusion to this book that all four versions are indispensable to an adequate evaluation of biotechnological enhancement in Christian ethics. In part, their indispensability is due to the fact that each version preserves something that is essential to the theology and ethics of human nature.

It does not follow, however, that all four versions are equally viable. The conclusion makes a modest case for the superiority of NS2 and NS4 over NS1 and NS3 and then for NS4 over NS2. More broadly, it states what I think Christian ethics can and should affirm about the normative status of human nature based on the analysis of Chapters 2 through 5; then it turns to the relationship of considerations involving the normative status of human nature to standard bioethical considerations involving autonomy, safety, and fairness. Because the enterprise of biotechnological enhancement is very much in flux, I refrain from discussing how all these considerations figure in the evaluation of particular biotechnologies. The relative priority of the various considerations and the weight given to each of them would vary depending on the circumstances surrounding these particular technologies, which cannot be determined in advance..

Finally, I deliberately relegate to an appendix a special matter that many people consider to be the most immediate and urgent concern raised by human biotechnology. Human biotechnology holds out the prospect that sudden or accumulated changes to human biological functions and traits may pass a threshold beyond which those have undergone them will have become something other than human. So-called transhumanists embrace this prospect, or at least remain sanguine in the face of it, while some of their critics contemplate it with alarm. These proponents and critics of transhumanism share the assumption that this prospect is an urgent one that demands our immediate attention. However, that assumption is a questionable one, and there are good reasons not to focus too quickly or too narrowly on this prospect. For one thing, it is doubtful that we have a sufficiently determinate concept of human nature to be able to identify in advance the point at which human beings become something else. Furthermore, this prospect is in any case a distant and uncertain one, and preoccupation with it carries the risk that we will be insufficiently attentive to more immediate and urgent matters. There are good reasons, then, for not prioritizing this issue in a treatment of the normative status of human nature. However, no position that attaches normative status to human nature can avoid the question whether the succession of human nature by some other kind of nature would be consistent with this status or not. The point of the appendix follows from Chapter 5. I argue that the present characteristics and capacities of humans are those in and through which they enjoy the particular life with God and other humans for which humans were created – a life that might not be available to those who are no longer human.

Normative Status of Human Nature: Its Meaning

To repeat it one more time, this book carries out what I take to be a necessary task of Christian ethics in relation to biotechnology, namely, to determine what the normative status of human nature is and how it counts in the evaluation of biotechnological interventions that implicate human nature, and it carries out this task by critically reconstructing and evaluating the four positions I have just described. But what exactly does it mean to claim that normative status attaches to human nature? The four positions I have just described assert that human nature has some *normative significance* that is relevant to biotechnology. This normative significance has to do (though not exclusively or reductively) with the biological aspect of human nature as a constituent of the created order or the human person; as a ground of human rights or goods, of personal identity, or of agency; or as a bearer of the meaning or purpose for which God created human beings. The four positions are versions of the claim that normative status attaches to human nature based on, and with respect to, its normative significance. But what exactly does that claim mean?

To say that *normative status* attaches to human nature is fundamentally to say that the implications of biotechnological interventions for human nature should count in the ethical evaluation of those interventions. They must be taken into account if the ethical evaluation is to be sound. Any particular version of this claim would also, if adequately developed, include claims about *how* these implications count, identifying which biotechnological interventions they apply to and elucidating how they apply to them. And finally, a complete account of this claim would also go on to say how much these implications for human nature should count relative to other morally significant factors, including autonomy, safety, and fairness. With respect to this last point, it is important to note that to claim that human nature counts in the evaluation of interventions that implicate it is not necessarily to claim that it overrides all other relevant factors. For example, one may consistently argue both that it is ethically questionable to extend the biological life span *and* that it is unjustifiable to interfere with the choices of individuals to pursue greater longevity, just as one may consistently argue both that a broad duty to promote human perfection requires cognitive enhancement *and* that considerations of safety or fairness will for the foreseeable future override that duty. In ethically evaluating biotechnological enhancement, none of the positions treated in this book would defend coercion or overlook safety or fairness. However, all of these positions at least imply that the normative status of human nature should count in some fundamental and significant way in the ethical evaluation of biotechnologies that implicate human nature.

To say that normative status *attaches to* human nature is to keep in play a wide range of positions. This range extends from claims that normative status inheres in human biological nature (for example, on the grounds that human biological functions and traits in their present form fully instantiate God's purpose regarding creation) to claims that human nature has no normative status unless and until we confer it (for example, on the grounds that we would violate certain moral obligations to someone, such as the obligation to respect her autonomy or to love her unconditionally, if we were to determine her biological nature). The first position is an exceedingly strong one, and many people find it neither plausible nor attractive. The second position is not implausible, and it receives attention in this book, but most Christian authors, at any rate, have something stronger in mind when they defend the normative status of human nature. According to most of the positions I treat in this book, the claim that normative status attaches to human nature falls somewhere between these two extremes and involves the claim that human nature is a constituent of something else (for example, the human person, the order created by God, or the *imago dei* which humans are or reflect) that is the proper bearer of normative status. According to these positions, human biological nature enjoys its normative status neither in itself nor simply by our good pleasure in conferring it but as an essential and irreducible aspect of, say, the human person, the created order, or the *imago dei*. However, I want to make room for the full range of positions that have been taken in debates over biotechnology, including the implausibly strong and overly weak positions. The term *attaches to* accommodates that range, and it is the term I use in what follows.

Finally, to say that normative status attaches to *human nature* is, for purposes of this book, to say that it attaches to human nature with respect to some feature(s) the latter has or some role it plays. For the most part, authors who claim that normative status attaches to human nature do not begin with a precise definition, a comprehensive concept, or an ideal description of human nature and then go on to argue that biotechnology should protect or promote human nature, so understood. Rather, on the basis of their views of the normative significance of human nature, they pick out one or more features of human nature or some role that it plays and attach normative status to human nature by virtue of or with respect to the relevant feature(s) or role.[6] This is an important point to keep in mind in what follows. My examinations of the positions treated in this book focus primarily on the feature(s) or role to which normative status attaches and on what follows for biotechnological actions that implicate the relevant feature(s) or role; they do not for the most part focus on definitions, descriptions, or concepts of human nature as such.[7] Indeed, insofar as ascriptions of normative status do not directly pertain to human nature as such but to determinate feature(s) or roles of human nature, it would be a distraction to focus on human nature as such.

If, then, I and the authors I treat may be exempted from the requirement to provide a theory of human nature as such, we are not exempted from thinking carefully about human nature insofar as normative status attaches to it. To be precise, it is necessary and important to the task at hand (1) to define and describe the relevant feature(s) or roles of human nature to which normative status attaches and (2) to determine whether human nature genuinely possesses the feature(s) or plays the role attributed to it. With regard to the definition and description of features or roles, I have mentioned that the positions treated in this book find normative significance in some meaning or purpose that human nature has or in human nature as a ground of goods, rights, identity, or agency or as a constituent of the created order or the human person. This book shows how these bases of normative significance determine which feature(s) or role of human nature normative status will attach to. For example, some subscribers to NS1 view human biological nature as normatively significant insofar as the inviolability of the person extends to it, and I show how they accordingly attach normative status to human biological nature as that which is undetermined by intentional human action. The normative significance of human biological nature is found in its inviolability as a constituent of the person or as a condition of identity or agency, and normative status therefore attaches to human biological nature as that which stands over against intentional human action. Meanwhile, for some subscribers to NS3, human biological nature is the material for a distinctively

---

[6] Gregory Kaebnick makes this observation with respect to Fukuyama, Habermas, Kass, Murray, Nussbaum, and Sandel. See Kaebnick, "Human Nature without Theory," in Kaebnick, ed., *The Ideal of Nature*, pp. 49–70.

[7] As the appendix makes clear, one context in which the normative status of human nature simply as such is at issue is the prospect of the "posthuman," in which biotechnological alterations amount to a change of the nature of at least some humans into something else.

human vocation to continue or perfect God's work of creation. Its norma-tive significance is found in this purpose, and I show how NS3 accordingly attaches normative status to human nature as indeterminate, open-ended, and malleable, and thus as susceptible to human intervention.

With regard to the determination of whether human nature actually possesses the relevant feature(s) or plays the relevant role that is picked out by the ascription of normative status, the minimum condition of the plausibility of a claim that normative status attaches to human nature is that the latter possess the relevant feature or the capability for the role to which normative status attaches. In the two examples just given, the task, respectively, will be to determine whether we can plausibly regard human nature as in some sense left undetermined by intentional human action or as in some sense and to some extent susceptible to human intervention. If (as many believe) there isn't any sense in which human nature is undeter-mined by intentional human action, then NS1 will lack the minimum con-dition for attaching normative status to human nature as that which stands over against intentional human action. Similarly, if human nature turns out to be recalcitrant to biotechnological intervention, then NS3 will lack the minimum condition for attaching normative status to human nature as indeterminate, open-ended, and malleable.

Finally, in what follows I will generally speak of "the normative status of human nature" rather than "the normative status of human biological nature." It is true that the latter term is more precise, given the focus of this book. Because this book is concerned with the implications of bio-technological enhancement for human nature, it focuses on the biological aspect of human nature. At the same time, however, "human biological nature" might suggest a narrow or reductionist view that I want to avoid. Biotechnological enhancement implicates biological functions and traits to bring about desired characteristics, capacities, or states. As such, it dir-ectly implicates human biological nature, but the normative relevance of what it does is not limited to the biological functions and traits that it directly implicates; rather, it extends to the entire range of human charac-teristics, capacities, and states that can be affected by these functions and traits. In other words, every aspect of human nature is at least potentially at stake in the capability of biotechnology to implicate human biological nature even when claims about the normative status of human nature are explicitly about biological nature. These considerations favor the use of the term *the normative status of human nature* except in contexts where clarity is served by using the term *the normative status of human biological nature.*

Problems with Normative Discourses on Human Nature

Claims that normative status attaches to human nature are subject to both moral and conceptual criticisms. These criticisms are exceedingly serious, and no inquiry into the normative status of human nature can ignore or marginalize them. The criticisms have been pressed by two very different groups, namely, mainstream bioethicists (the most important of whom, for purposes related to our topic, are philosophers), on the one hand, and theorists and scholars in the fields of gender studies, critical race theory, queer theory, disability studies, and science and technology studies, on the other hand. I will consider the criticisms of these two groups separately, beginning with those of the bioethicists.

Many mainstream bioethicists dismiss the normative status of human nature on the grounds that it involves conceptual confusion. For these bioethicists, human nature is simply what results from and operates according to biological processes. To attach normative status to human nature so understood is to commit a kind of category mistake. Frances Kamm puts the point succinctly: "The natural and the good are distinct conceptual categories and the two can diverge."[8] It is clear that much that results from biological processes, including diseases and antisocial behaviors, is plainly not good, while many things that do not result from those processes, such as eyeglasses and pacemakers, are plainly good. Of course, the reverse is also true; what is natural is often good and what is not is sometimes bad. The point is that the natural and the good are distinct normative categories, not that they never coincide. Bonnie Steinbock draws the conclusion that follows from this point: "That something is natural is not in itself a reason for thinking it is good; that it is unnatural is not a reason for thinking it is bad."[9] From this perspective, to attach normative status to human nature is to invest with normative significance what is in itself indifferent to what we recognize as good or valuable.

This line of criticism makes two assumptions about claims that normative status attaches to human nature. First, it assumes that the human nature to which normative status allegedly attaches is no more than biological processes and whatever results from them, and second, it assumes that the normative status that allegedly attaches to human nature inheres in human

---

[8] Frances Kamm, "What Is and Is Not Wrong with Enhancement?" in *Human Enhancement*, edited by Julian Savulescu and Nick Bostrom (Oxford: Oxford University Press, 2009), p. 103.

[9] Bonnie Steinbock, "The Appeal to Nature," in *The Idea of Nature: Debates about Biotechnology and the Environment*, edited by Gregory E. Kaebnick (Baltimore: The Johns Hopkins University Press, 2011), p. 110.

nature so understood. However, theologians and philosophers who attach normative status to human nature reject one or both of these assumptions. Some of them (including Habermas) accept the first assumption but deny that normative status inheres in human biological nature, so understood, arguing instead that it inheres in something else (in Habermas's case, the human person) of which biological nature is a constituent. Others reject the first assumption, arguing that human nature insofar as normative status attaches to it includes principles of order or purpose that, even if they are at least partly intelligible in terms of biological processes, are not simply read off those processes or reducible to them. If we have reason to reject either or both of these assumptions, then we are justified in holding that mainstream bioethicists have not successfully made their case against the claim that normative status attaches to human nature.[10]

Scholars and theorists of gender, race, sexuality, disability, and science and technology have uncovered both problematic assumptions about human nature that underwrite many normative discourses on human nature and disturbing moral legacies of these discourses, which have been especially pernicious in the realms of gender, sexuality, race and ethnicity, and disability. These conceptual and moral problems attend discourses on human nature that feature one or more of the following: (1) a concept of human biological nature as unchanging and/or clearly demarcated from the nonhuman (whether animal or machine), which grounds essentialist claims about human nature; (2) a separation of human biological nature from society and culture, which isolates claims about human nature from social and cultural constructions of it and thereby conceals social and cultural biases and hegemonies that discourses about human nature perpetuate and shields them from interrogation and criticism; and (3) the privileging of certain states or conditions of human nature as normative, on the basis of which other states and conditions are judged to be inferior, abnormal, or impure. In addition, some of these scholars join with philosophically trained bioethicists in pointing out an additional problem, namely, (4) the derivation of norms of behavior from descriptions of human nature.

There are many poignant criticisms of these four features, but my understanding of the problems that attend the first three of them has been especially influenced by three authors. With regard to (1), Donna Haraway has drawn attention to the indeterminacy and instability of the boundaries

---

[10] A notable exception is Allen Buchanan, whose arguments against the normative status of human nature are considered (mostly in footnotes) in the following chapters. See Buchanan, *Beyond Humanity?*

between humans, on the one hand, and animals and machines, on the other hand, and has explored the social and political implications that follow when the fluidity of these boundaries is taken seriously.[11] With regard to (2), Judith Butler has demonstrated that there is no access to human biological nature apart from constructions without which we can neither think nor make sense of our biological nature and has subjected constructions of gender and sexuality to analysis and critique.[12] With regard to (3), Lennard Davis has pointed out how determinations of human functions and traits in terms of their conformity to or deviation from norms that represent ideal or optimal states is a recent phenomenon that owes more to the interests of modern societies in the management of bodies than it does to the scientific objectivity in which the determinations are couched.[13]

At certain points in the following chapters, criticisms along the lines of those that have been made by these authors will be explicitly formulated; at other points they will implicitly shape my presentation of claims about the normative status of human nature. However, it is worth asking at the outset whether *any* normative discourse on human nature, including this one, can survive the critiques of features (1)–(4) that have been proffered by Butler, Davis, Haraway, and others. Would not any claim that normative status attaches to human nature in the context of biotechnology run afoul of at least one of these critiques? And considering the disturbing moral legacies that these critiques have exposed, would it not be best to avoid normative discourses on human nature altogether?

I will reserve my answer to the second of these questions for the following section, but two points may be made in response to the first question. In the first place, these critiques *do* impose significant constraints on any claim that normative status attaches to human nature. If, as the critics assert, human nature lacks features (1) and (3), so that it cannot be definitively demarcated from the nature of other animals or from machines, and it does not conform to metaphysical or biostatistical norms, then normative status cannot attach to it with respect to characteristics or roles that presuppose (1) and (3). And if (2) and (4) are illegitimate moves, so that attempts to represent human nature apart from constructions and

---

[11] See especially Donna J. Haraway, *Simians, Cyborgs, and Women: The Reinvention of Nature* (New York: Routledge, 1991).

[12] See Judith Butler, *Bodies That Matter* (New York: Routledge, 1993). See also Donna J. Haraway, *Modest_Witness @ Second_ Millenium. FemaleMan_Meets_Oncomouse: Feminism and Technoscience* (New York: Routledge, 1997), pp. 49–118.

[13] See Lennard J. Davis, *Enforcing Normalcy: Disability, Deafness, and the Body* (London: Verso, 1995); and Davis, "Introduction: Disability, Normality, and Power," in Lennard J. Davis, editor, *The Disability Studies Reader*, 4th ed. (New York: Routledge, 2013), pp. 1–14.

to derive behavioral norms from descriptions of human nature are illicit, then claims that normative status attaches to human nature will have to avoid those moves. These constraints clearly rule out many versions of the claim that normative status attaches to human nature in the context of biotechnology.

In the second place, however, there is no reason to assume at the outset that these constraints leave no room for *any* claim that normative status attaches to human nature. To begin with (1), the claim that the boundaries of human biological nature are fuzzy and shifting does not entail that there is no such thing as human nature or that the latter has no distinguishable feature(s) or role to which normative status can attach. It simply means that human nature does not have strict boundaries, and normative status cannot properly attach to any characteristic or role that implies that human nature does have such boundaries. Haraway provides a good example of how one might attach normative significance to human nature without running afoul of her critique when she affirms, at least in a qualified sense, the legitimacy of the notion that biological species (including the human species) are natural kinds (and thus stable enough for normative status to attach to them), and finds normative significance in the indeterminate character of human nature, the implications of which she explores for their emancipatory potential.[14] Turning to (2), Butler does not deny that there is such a thing as biological nature, and her claim that constructions of biological nature have a necessity that is no less compulsory than biological necessity lends support to the notion that normative discourses on human nature are unavoidable. Her critique does not render all such discourses illegitimate in principle but rather calls for vigorous interrogation of them

[14] For the first point, see Donna Haraway, *The Companion Species Manifesto: Dogs, People, and Significant Otherness* (Chicago: Prickly Paradigm Press, 2003), p. 15. For the second point, see Haraway, *Simians, Cyborgs, and Women*, pp. 148–92. Of course, many of those who, like Haraway, hold that contemporary technoscience has rendered implausible the notion of an intact, unchanging human nature that is clearly demarcated from the natures of other things infer from this circumstance (or simply assume) that there is now no such thing as human nature, if indeed there ever was such a thing. For some of these critics what is at stake in biotechnology is therefore not human nature – that is already out of the picture – but the human subject or self, and in particular whether the subject or self will be identified with disembodied patterns of information (cybernetic selfhood) or with finite embodiment (cyborg selfhood). See, for example, Katherine Hayles, *How We Became Posthuman: Virtual Bodies in Cybernetics, Literature, and Informatics* (Chicago: University of Chicago Press, 1999); and Jeanine Thweatt-Bates, *Cyborg Selves: A Theological Anthropology of the Posthuman* (Aldershot: Ashgate, 2012), both of whom affirm the superiority of the cyborg alternative over the cybernetic one. The debate over these alternatives is an important one, and I stand with Hayles and Thweatt-Bates in favoring the cyborg alternative over the cybernetic one. But because it is a debate about the subject or self rather than one about human nature, it falls outside the scope of this study.

and constant vigilance over them. Continuing with (3), we will see that both biological science and the philosophy and theology of human biological nature offer plausible alternatives to the portrayal of human biological functions and traits as conforming to or deviating from natural norms. If these alternatives prove to be viable, then the nature to which normative status attaches is not norm-governed in the pernicious sense which the critique targets. Turning finally to (4), there are, as we will see, a variety of plausible ways to attribute normative significance to human nature without deriving norms of behavior from it.

In the chapters that follow, I will be guided by three points regarding these problems. First, there is no reason to assume at the outset that any claim about the normative status of human nature will run afoul of the criticisms of Butler, Davis, Haraway, and other critical theorists. Second, it is nevertheless necessary both to preserve the gains in our understanding of human nature that their criticisms have made possible and to maintain vigilance in light of the dangers that attend every claim that normative status attaches to human nature. Third, it is worth considering that the most effective way to oppose problematic normative constructions of human nature might be to formulate explicitly normative conceptions that rule out problematic features rather than to suppose that these constructions can be effectively dismantled by simply deconstructing them or describing human nature more adequately. In other words, the best way to avoid the problems with normative conceptions of human nature might not be to avoid formulating normative conceptions of human nature but to try to formulate conceptions that avoid the problems. Of course, that is only a hunch, but if normative discourses on human nature are unavoidable in any case, it is a hunch that even those who have no investment in such discourses should consider worth testing.

In what follows, I will try to account for both the possibility of formulating claims about the normative status of human nature that do not in principle incur the problems with such claims and the dangers that in practice always accompany such claims in this way: First, wherever I encounter formulations of these claims that do run afoul of the criticisms of these features, I will show how the formulation is disqualified by this circumstance. Second, whenever possible, I will then go on to formulate the claim in a way that avoids the problematic feature. In the end, I am confident that the positions I treat in this book can be formulated in ways that avoid these problematic features, though I do not claim that either I or the authors I treat have adequately done so.

What Is at Stake for Christian Ethics?

If it turns out that claims that normative status attaches to human nature can avoid the four problematic features we have just considered, we may still wonder whether it is worth the effort to formulate such claims when we consider the pressing ethical issues that are raised by biomedical research and practice. In my view, four such issues are especially pressing. One is the provision of health care to the poor. Another involves race, gender, and sexuality as factors in the delivery of health care and the conduct of biomedical research. The third issue has to do with the destruction of embryos in biotechnology research and biomedical practice. The final issue involves physician-assisted suicide and euthanasia in the context of end-of-life care. For anyone who is struck by the gravity of these issues, attention to what biotechnology might do with regard to human nature seems to lack urgency and even importance. Does it really matter in comparison with these obviously pressing issues?

One answer to this question points to how the capability of intentional intervention into human nature poses profound challenges to some of the major concerns of Christian ethics. For example, until recently the inviolability of human biological nature (that is, its nonsusceptibility to intentional intervention) provided a last line of protection of persons from reduction to objects or things, or subjection to the desires and expectations of other persons. Thanks to inquiries in many fields in recent decades, we are now aware of how human bodies can be and are reduced and subjected in these ways, and how human subjects can be and are thereby formed and unformed, but until recently human biological nature stood as a limit to human power over and domination of other humans (even as representations of human nature were used in morally disturbing ways, as we have just seen). However, as proponents of NS1 emphasize, that limit has been breached, and we must now determine whether appeal to bioethical principles such as autonomy and safety is sufficient to secure the protection of persons or whether, on the contrary, it requires us to keep human nature off-limits to intentional actions to determine it. To take another example, until recently Christian ethics could assume that the goods that God created us to enjoy are goods that fulfill us as creatures of *this* nature: that is, the nature we now have, and not some other. It could also point to the ordering of our nature to our good as an indication of the goodness and intelligibility of the world as God's creation. But as debates over NS2 (which are detailed in Chapter 3) make clear, Christian ethics now face the question whether our nature should determine what we should recognize as our creaturely good or whether what we recognize as our good should determine what our nature shall be, and whether the mark of the goodness of the world as God's creation is not the ordering of our nature as it is to our good but rather the capability of our nature to become something other than it is. Finally, even when Christian ethics takes human biotechnology to be an ally, as it is for subscribers to NS3, its concerns are not immune against challenges from biotechnology to the extent that the actual course of development of biotechnology threatens to frustrate its emancipatory potential or reduce its visions of human perfection to the banality of bourgeois improvement. These considerations all suggest that what biotechnological enhancement does with respect to *our nature*, it does to *us*; and in doing so, it involves moral values that are central to Christian ethics.

Another answer to the question of the importance to Christian ethics of a serious investigation of the claim that normative status attaches to human nature points to the inevitability of normativity in human bio-technology. Every program of biotechnological enhancement inevitably instantiates *some* normative conception(s) of human nature, and given their potential impact, these programs could determine to a nonnegligible extent the values, purposes, and roles that will attach to human nature in the future. Human nature is no longer the "given" that supplies the condi-tions for intentional human action; it is now in play, so to speak, and the question of the values, purposes, and roles that human nature will embody, express, or serve will be answered one way or another – either by critical reflection on them or by default.[15] Because Christian ethics is invested in human nature, it has good reasons not to leave its normative significance to whatever the default position of biotechnological research and develop-ment turns out to be. The necessity of critical reflection would still hold even if the result of reflection turns out to be an endorsement of programs

---

[15] The likely default position is the one found in Julian Savulescu, Anders Sandberg, and Guy Kahane, "Well-Being and Enhancement," in Julian Savulescu, Ruud ter Meulen, and Guy Kahane, eds., *Enhancing Human Capacities* (Oxford: Wiley-Blackwell, 2009), pp. 3–18. This position features three claims, which are asserted on pp. 10–15: (1) that the human good is specific to individuals; (2) that biotechnological enhancement should focus on "all-purpose goods" that are conducive to any way of life one might choose; and (3) that the value of enhancements such as increased cogni-tive capability is due to their role in facilitating greater skill in instrumental reasoning so that one is better able to choose the means to one's ends, whatever they may be. These claims reflect liberal democratic values that are appropriate in their own place. But they also express, respectively, three key requisites of late capitalist economies, namely, the ephemerality of a purely individual good that is readily susceptible to manipulation by advertising and marketing; an optimally adaptable labor force that is responsive to abrupt shifts in labor markets; and maximum efficiency in means. Whatever Christian ethics takes to be the value or purpose of human biological nature, it will not be found in (and may well be incompatible with) these characteristics.

of human enhancement and of the normative conceptions these programs instantiate.

That intentional intervention into human nature is accomplished by way of technology lends urgency to the need for reflection on how human biotechnology instantiates normative conceptions of human nature. The kind of reflection I have in mind focuses on the role of human biotechnology in the medicalization and normalization of human nature. Medicalization is the process by which human biological functions and traits, and the characteristics and capacities they underwrite, become problems calling for biomedical intervention. Familiar examples of medicalization include the pathologization of socially disruptive behaviors, cognitive limitations, and social awkwardness, all of which are targets of psychopharmalogical interventions. In the course of medicalization, the relevant characteristics and capacities are subjected to criteria of measurability, predictability, and reproducibility and evaluated with respect to their conformity to or deviance from norms that reflect societal interests in an optimally efficient, adaptable, and productive population – a process known as normalization. Of course, to point out how biotechnology medicalizes and normalizes human nature is not to answer any normative questions about biotechnology. It may turn out that there are compelling ethical reasons for endorsing or at least not opposing the state of affairs that medicalization and normalization bring about. But in aiming at efficiency, adaptability, and productivity, medicalization and normalization instantiate certain normative conceptions of human nature, and in light of their pervasive and problematic effects it is worth asking whether those conceptions can be endorsed or must be modified, qualified, or rejected by Christian ethics.

Finally, as our nature is created by God, Christian ethics must, as a requirement of respect for God as Creator and for what God has created, ask what is the point (that is, the value or purpose) of our biological functions and traits and the characteristics and capacities they underwrite and whether biotechnological enhancements instantiate or violate that point. Upon consideration, those enhancements may finally be welcomed as instantiations or rejected as violations of the point of our biological nature. But whichever answer is ultimately given, the question of the point of our nature can no longer default to a nature that biotechnology does not implicate. Because biotechnology has put human nature in play, if only in aim and ambition, the question of the point of human nature must now be posed in light of biotechnology. That is what the following four chapters will now go on to do.

CHAPTER 2

Human Nature as Given

According to a highly influential position in debates over biomedical enhancement technology, normative status attaches to human nature as that which exists, and is what it is, apart from our willful activity. This conception of the normative status of human nature, which I will refer to as NS1, draws on the familiar idea of nature as a realm whose objects and processes owe their existence and characteristics to forces that operate independently of intentional human action. In the words of Oliver O'Donovan, nature is "a world which we have not made or imagined, but which simply confronts us."[1]

Although our concern is with human nature, it is notable that this idea of nature, and the claim that normative status attaches to nature so defined, appears in multiple contexts. One context involves environmental ethics and policy. The US Wilderness Act of 1964 famously defines *wilderness* as "an area where the earth and its community of life are untrammeled by man, where man himself is a visitor who does not remain."[2] Along the same lines, the philosopher Bernard Williams characterizes nature as that which "is not controlled, shaped or willed by us, a nature which, as against culture, can be thought of as just *there*."[3] The same idea is sometimes invoked in debates over transgenic organisms. Biologist Stuart Newman, a critic of research involving the transfer of genes between species, distinguishes these organisms from nature as "a world neither made nor influenced by human activity."[4] According to this conception, to intervene into nature

---

[1] Oliver O'Donovan, *Begotten or Made?* (Oxford: Clarendon Press, 1984), p. 3.
[2] Wilderness Act (Public Law 88–577) (16 U.S. C. 1131–1136), Section 2(c).
[3] Bernard Williams, "Must a Concern for the Environment Be Centred on Human Beings?," in *Making Sense of Humanity and Other Philosophical Papers* (Cambridge: Cambridge University Press, 1995), p. 240.
[4] Stuart A. Newman, "Renatured Biology: Getting Past Postmodernism in the Life Sciences," in *Without Nature? A New Condition for Theology,* edited by David Albertson and Cabell King (New York: Fordham University Press, 2010), p. 102.

at all is in effect to destroy it: Nature as controlled or altered is no longer that which is simply given or that "which we have not made or imagined, but which simply confronts us." At the moment of intervention, what was previously natural becomes an artifact. If it is to remain part of nature, then, it must be left as it is. And (to return to our main theme) if normative status attaches to human nature, so understood, then in principle any attempt to alter, control, replace, or select human biological characteristics will violate that status.

To assert NS1, then, is to declare that human nature should be kept off-limits to attempts to alter, control, or otherwise determine it. According to NS1, we should not try to extend the human life span, increase our cognitive capabilities, narrow the range of our emotions (for example, by making ourselves less aggressive or shy), or do anything else that would intervene into our biological functions or traits. The appeal of NS1 to critics of biotechnological enhancement is obvious: To declare human nature off-limits to intervention is the surest way to oppose the alteration and control of human functions and traits. In an earlier generation, Hannah Arendt and Hans Jonas expressed their opposition to human biotechnology in the idiom of NS1.[5] More recently, Jürgen Habermas, Oliver O'Donovan, and Michael Sandel have done so, and their arguments are the focus of this chapter.[6] These arguments have been subjected to severe criticism by mainstream bioethicists and moral philosophers, but the critics have sometimes misunderstood the arguments, while for their part the defenders have not always presented them in their strongest forms. The task of this chapter is to formulate and critically examine what I think are the strongest forms of the arguments made by these three subscribers to NS1. I will conclude that NS1 is not a defensible position on the normative status of human nature, but that it identifies considerations that should count in the ethical evaluation of biotechnology.

---

[5] Hannah Arendt, *The Human Condition* (Chicago: University of Chicago Press, 1958); Hans Jonas, *The Imperative of Responsibility: In Search of an Ethics for a Technological Age* (Chicago: University of Chicago Press, 1984).

[6] The following discussion will refer to Jürgen Habermas, "The Debate on the Ethical Self-Understanding of the Species," in *The Future of Human Nature* (Cambridge: Polity Press, 2003), pp. 16–100; Oliver O'Donovan, *Begotten or Made?*; idem, *Resurrection and Moral Order: An Outline of Evangelical Ethics*, 2nd ed. (Grand Rapids, MI: Eerdmans, 1994); Michael J. Sandel, *The Case Against Perfection: Ethics in the Age of Genetic Engineering* (Cambridge, MA: Harvard University Press, 2007); idem, "Mastery and Hubris in Judaism: What's Wrong with Playing God?," in Sandel, *Public Philosophy: Essays on Morality in Politics* (Cambridge, MA: Harvard University Press, 2005), pp. 196–210 (originally published in *Judaism and Modernity: The Religious Philosophy of David Hartman*, edited by Jonathan W. Malino [Aldershot: Ashgate, 2004], pp. 121–32).

Before turning to the arguments of O'Donovan, Habermas, and Sandel, I will consider four objections to NS1. Two of these objections come primarily from mainstream bioethics, the other two from critical studies of science and technology. I have two reasons for beginning with these objections. First, if any one of the objections turns out to be valid, then NS1 is disqualified from the outset as a viable position and there is no reason to go on to examine particular versions of it. It makes sense, then, to begin with the objections so that we can determine up front whether NS1 is a viable position. As we will see, I think that NS1 can meet each of these objections. The problems that in my view render NS1 indefensible in the end lie elsewhere, as the treatments of O'Donovan, Habermas, and Sandel will make clear. However, the objections identify problems that any version of NS1 will have to avoid if it is to be plausible at all. This brings me to my second reason for beginning with these objections. Working through them will enable us to refine the fundamental claims of NS1 so that the formulation of NS1 that emerges is a plausible one that avoids the most obvious errors that accompany attempts to talk about human nature as that which exists and is what it is apart from intentional human action.

The chapter therefore proceeds as follows. In the first section, I introduce the four objections, showing how NS1 can meet them and how they lead us to a plausible formulation of NS1. The following three sections, in order, formulate and critically examine the versions of NS1 that have been articulated by O'Donovan, Habermas, and Sandel. The chapter then concludes with my reasons for rejecting NS1 while also affirming its central concerns.

Before turning to the four initial objections, one preliminary point must be noted. Many defenders of NS1, including Habermas and Sandel, focus on genetic technologies. They worry about the use of genetic engineering to design humans with desired characteristics and the use of preimplantation genetic diagnosis (PGD) to select embryos with those characteristics before implanting them in the womb. Because neither of these techniques is presently capable of securing the desired positive characteristics (although PGD can identify dispositions to numerous diseases), some people think that the alteration or selection of genes does not deserve the attention that has been paid to it by (among others) subscribers to NS1. However, it is not unreasonable to expect that these or some other genetic technologies will succeed in the future, and that prospect is plausible enough to justify the attention paid to these issues in this chapter.

Four Initial Objections to NS1

For many readers, NS1 might seem too implausible to merit our attention. Its interest in keeping human nature off-limits to biotechnology seems to be motivated by a conviction that human nature is sacred and therefore untouchable. Moreover, it seems to rule out all technological interventions whatsoever, including those that aim at the prevention and treatment of disease and injury. It also seems to require us to think of human nature as something that is just "there" ("a world we have not made or imagined"), when in fact we do not have access to human nature independently of culture and human nature is partly the product of our activity. These suspicions can be formulated as objections, any one of which would invalidate NS1.

The first initial objection to NS1 is partly based on a misunderstanding. Some bioethicists find NS1 unattractive because they assume that it involves an unexamined reverence for what is natural (that is, what is allegedly untouched by human activity) or a misplaced ascription of value to it. They are confident that once we question the assumption that nature is sacred and begin weighing whatever value it might have against the potential benefits of altering or controlling it, the appeal of NS1 quickly dissipates. It is true that some subscribers to NS1, including O'Donovan and Sandel, argue (as we will see) for keeping human biological nature off-limits partly because intervention into it involves attitudes or stances toward nature that reduce the latter to a mere instrument or an object of mastery. To be sure, it is not always apparent what is wrong with these attitudes or stances. Is an attitude or stance that instrumentalizes nature or exercises mastery over it problematic because of some sacredness or value that nature has, or is it problematic because there is something wrong with the attitude or stance? In either case, however, this first objection is a relevant one, and we will see when we get to O'Donovan and Sandel whether it is valid. It is important, however, to emphasize that the strongest and most influential arguments of subscribers to NS1 are not subject to this objection. Habermas, O'Donovan, and Sandel have all put forth arguments that do not attach normative status to human nature on the grounds that what is natural is sacred and thus untouchable or on the grounds that its value is greater than any value that could be realized by altering, selecting, or controlling it; nor do these arguments focus on attitudes or stances to nature. Rather, they echo, in their distinctive voices, a worry expressed by C. S. Lewis in *The Abolition of Man*. To take intentional control of human nature, Lewis famously asserts, is to exercise power over other human beings: "[W]hat we call man's power over nature turns out to be a power exercised by some men over other men with nature as its instrument."[7] The most plausible versions of NS1 argue that biotechnological alteration, selection, or control of biological characteristics involves the illicit power of some human beings (usually parents) over other human beings (usually children), and in attaching normative status to human nature as that which is simply given, these versions seek to protect vulnerable human beings from the power of others. The worry about illicit power explains why NS1 continues to be attractive to many people, if not to most bioethicists. One aim of this chapter is to determine, by critically examining these arguments, whether that worry is warranted in the case of genetic technologies.

The second initial objection against NS1 also comes from mainstream bioethics. The proposal that we should leave human nature as it is, appears to rule out interventions to prevent or cure diseases and injuries. If we can respect the normative status of human nature only by letting it be in its sheer "givenness," must we not simply acquiesce to such threats when they arise, leaving nature to itself? In response to this objection, subscribers to NS1 distinguish between treating a disease or injury, on the one hand, and enhancing a trait or function, on the other hand. The prevention and cure of diseases and injuries, they argue, aim at the preservation or restoration of the nature of the organism, not at its alteration, even if enhancement of functions or traits may result from some efforts at prevention or cure (as, for example, vaccination to prevent a communicable disease enhances the immune system). From this perspective, the nature of the organism is not simply its brute otherness but rather has to do with functions and traits in an integrated whole. To intervene to maintain or restore this integrated whole from a threat to it is therefore to leave the nature of the organism as it is, and this can be distinguished, at least in principle, from interventions that select, replace, or alter functions and traits. However, this argument on behalf of NS1 assumes that therapy and enhancement can be distinguished from one another. This is a highly controversial assumption. To consider it adequately would require a long detour through the philosophy of biology and medicine, where it has been widely debated.[8] Because it would be disproportionate to the significance of the issue for this chapter, I will avoid the detour and simply register my suspicion that those who hold

---

[7] C. S. Lewis, *The Abolition of Man* (London: Macmillan, 1947), p. 69.
[8] The debate over this issue forms a large body of literature. For a recent and thorough critical discussion of the literature, see Neil Messer, *Flourishing: Health, Disease, and Bioethics in Theological Perspective* (Grand Rapids, MI: Eerdmans, 2013).

that the distinction cannot finally be upheld either conceptually or practically are closer to the truth. If that is so, NS1 collapses due to its inability
to rigorously distinguish therapeutic interventions that respect nature as
given from enhancement interventions that violate its givenness. However,
because the debate is inconclusive, I will concede the viability of the distinction and give NS1 a pass on this objection. Meanwhile, consideration
of the objection has allowed us to refine the fundamental claim of NS1 in
an important way. Nature as that which exists and is what it is apart from
our intentional action is not the brute over-against-ness described by the
Wilderness Act, Williams, and Newman, but rather, at least in the case of
the human organism, an integrated whole of functions and traits. To leave
nature as it is, is therefore not to rescind from all intervention whatsoever,
but rather allows for those that maintain or restore the integrated whole.

The third initial objection to NS1 comes from the critical study of science
and technology. It charges that descriptions of nature, including human
biological nature, as "a world which we have not made or imagined" or
"which simply confronts us" are both naïve and morally problematic. It is
now clear both that our access to human biological nature is always at least
in part mediated by culture and that the ideal of a pure human biological
nature that is independent of culture has been (and perhaps is inevitably)
deployed in morally repugnant ways.[9] These criticisms are well-founded
and morally weighty. We have no access to our nature that is not mediated through language, practices, and institutions. Of course, it does not
follow that nature is merely an aspect of culture or that what we refer to
as nature is nothing other than culture. But it does mean that what we
represent as natural in the cases of traits (such as longevity) and functions
(such as cognitive performances or emotional proclivities) reflects the ideals and the actual influences of particular societal and cultural conditions
and not just nature. With respect to our nature, then, there is no world
that we have not imagined, or which simply confronts us. Meanwhile,
history supplies ample evidence that conceptions of what is ideal, pure, or
normative with respect to our nature, along with imperatives to protect
it, are constructions that can be perilous for those who fail to conform
to them. These considerations are unassailable, and they make this third
objection a formidable one. However, NS1 can be articulated in ways that
avoid this objection. What emerges is, once again, a more plausible formulation of the fundamental claim of NS1. My formulations of the versions
of NS1 proffered by O'Donovan, Habermas, and Sandel will make clear

---

[9] These points are forcefully made in Judith Butler, *Bodies That Matter* (New York: Routledge, 1993).

that the claim that human traits or functions should be kept off-limits to biotechnological intervention is compatible with the recognition that our theoretical and practical engagements with human nature are mediated by cultural constructions, and that it need not assume that there is any ideal, pure, or normative state or condition of human nature at all, much less that there is an obligation to preserve such a state or condition. The claim is simply that we should not intentionally alter or select the biological characteristics of others. That claim is compatible with the claim that we have no access to our biological nature apart from social constructions, and it makes no assumptions about which biological states or conditions are ideal, pure, or normative, or even desirable or preferable.

The fourth initial objection to NS1 is the most challenging one. It is by now a well-established fact that human traits and functions have been significantly altered *by human activity*. As Hans Jonas argued a generation ago, our ancestors may have been justified in supposing that nature, including human biological nature, is the unalterable background of human action, but we cannot avoid reckoning with the reality that nature, including human biological nature, has become the object and even, to a significant extent, the product of human activity.[10] This point could easily be made by considering the evolutionary history of humans, but for our purposes it is more relevant to point to three undisputed examples that relate to commonly proposed biotechnological enhancements. One example is found in studies that have demonstrated that literacy enhances some functions of the brain (while possibly inhibiting other functions).[11] The other two examples are more familiar. In the United States, life expectancy at birth increased from 47.3 years to 76.9 years over the course of the twentieth century, while the average adult height of native-born men in the United States increased roughly three inches (from 66.9 to 69.8) during the first half of the twentieth century.[12] While the causes of the dramatic changes in longevity and height are widely debated, human activity undoubtedly played a significant role in them. If human nature is "a world which we have not made ..., but which simply confronts us," then it seems that it disappeared long ago, if indeed it ever existed.

---

[10] Jonas, *The Imperative of Responsibility*, pp. 1–23.

[11] For a recent study, see Stanislas Dehaene et al., "How Learning to Read Changes the Cortical Networks for Vision and Language," *Science* 330 (2010): 359–64.

[12] On longevity, see Centers for Disease Control, *National Vital Statistics Reports* 61.3 (2012), p. 52. On height, see Richard H. Steckel, "Health, Nutrition and Physical Well-Being," in *Historical Statistics of the United States: Millennial Edition*, vol. 2, edited by Susan Carter, Scott Gartner, Michael Haines, Alan Olmstead, Richard Sutch, and Gavin Wright (New York: Cambridge University Press, 2002), pp. 499–620.

This objection is a serious one. If human nature lacks the characteristic identified by O'Donovan (namely, that it is not made by us, but simply confronts us), then normative status cannot attach to it with respect to that characteristic. However, NS1 is not necessarily disqualified by this consideration because normative status may attach to human nature not as that which is unchanging or unsusceptible of human influence but as that which stands over against *intentional* human activity. According to this formulation, what is normatively significant is not some allegedly untouched state of our biological nature, but rather that our biological nature stands over against our willful activity. In this case, to declare our nature off-limits to deliberate intervention would not be inconsistent in principle.

Thus, one line of defense of NS1 against this fourth objection might stress that it is still plausible to attach normative status to human nature as that which is given because the changes such as those I just mentioned did not result from interventions that *aimed* at altering human biological functions and traits. This is, to be sure, a debatable point, and it also depends on the viability of the distinction between therapy and enhancement, which I have already noted is a precarious one. But if we allow that distinction, a defender of NS1 could argue that the best way to explain the programs in public health, nutrition, and so on that contributed heavily to the increases in life expectancy and average height is to hold that they did not, strictly speaking, aim at altering human nature but rather at health, broadly understood, while the changes literacy effected on the brain resulted from the exercise of an already existing capacity. In both cases, changes to biological functions and traits occurred as the by-product of service to or exercise of functions or traits of the human organism in their present condition, not by aiming at enhanced conditions. If this is so, then the characteristic of human nature to which normative status attaches, namely, its givenness, exists if (to amend O'Donovan's description) we think of human nature as "a world which we have *not intentionally* made, but which simply confronts *our willful activity*." This formulation preserves what really matters to the most plausible versions of NS1 because (according to these versions) it is only when we aim at alteration or control of the biological characteristics of other humans (as distinct from aiming at their health) that we exercise illicit power over them. While he does not explicitly state it, this defense is consistent with Jonas's version of NS1 (and perhaps also with Arendt's version).

A second (and more plausible) line of defense of NS1 against the fourth objection is less dependent on the debatable assumption that human nature possesses the characteristic (namely, givenness in the face of intentional human activity) to which normative status attaches. One version of this second line of defense (namely, O'Donovan's) holds that givenness is a characteristic of the created order itself and not of the things (including human biological functions and traits) that are ordered by it. Unlike the created order, the things that are ordered by it may indeed be, at least in part, the products of intentional human activity. Biotechnological determination of human biological nature is wrong on this view to the extent that it violates the given order of creation, and it is suspect to the extent that it takes up a stance toward things that disregards the order in which they exist. Other versions of NS1 concede that human nature has been altered by our intentional activity and is now at least in part under our willful control but view their task as one of restoring to it on the moral level a condition of immunity against human intervention that it has lost on the scientific and technological levels. They are attempts at what Wolfgang van den Daele has called "moralizing human nature," which occurs when "that which science made technologically manipulable reacquires, from a normative perspective, its character as something we may not control."[13] Immunity against intentional alteration is now a status *we must confer on human nature*; it is not (or is no longer) a characteristic of human nature itself. Jürgen Habermas's use of the German word *unverfügbar* nicely captures this point. Human nature, although it is now (at least to some extent) under our power, should not be put at our disposal or made available to us.[14] Because the English expressions "not at our disposal" or "unavailable" are cumbersome or imprecise, I use the term *immunity* as the equivalent of *Unverfügbarkeit*. According to this version of the second line of defense against the fourth objection, the claim of NS1 would read as follows: *Despite the fact that they may have come under our power, we should not consider human biological functions and traits to be at our disposal, and when we take them to be at our disposal by intentionally altering, selecting, or controlling them, we exercise illicit power over those whose functions and traits we alter, select, or control.* Two prominent versions of NS1 that assert this claim, even if not explicitly, are those of Habermas and Sandel. The rest of

[13] Wolfgang van den Daele, "Die Natürlichkeit des Menschen als Kriterium und Schranke technischer Eingriffe," *Wechsel-Wirkung*, June–August 2000, pp. 24–31, quoted in Jürgen Habermas, "The Debate on the Ethical Self-Understanding of the Species," in *The Future of Human Nature* (Malden, MA: Polity Press, 2003), p. 24. Van den Daele goes on to identify moralization with a resacralization of nature, but for the authors I examine in this book the normative status of nature is strictly moral and not sacral. The task of "moralizing human nature" may be said to have originated with Hans Jonas. See Jonas, *The Imperative of Responsibility.*

[14] Habermas, "The Debate on the Ethical Self-Understanding of the Species," p. 22.

this chapter is devoted to those who exemplify these two versions of the second line of defense against the fourth objection, namely, O'Donovan, who exhibits the first version, and Habermas and Sandel, who exhibit the second.

We began with the suspicion that the fundamental claim of NS1 – that human nature exists and is what it is apart from intentional human action and should be kept off-limits to biotechnology – implies that human nature is sacred; precludes all interventions whatsoever, including therapeutic ones; separates human nature from social and cultural constructions of it; and ignores the role of human activity in producing human nature. The NS1 that emerges from these objections holds that the givenness of our nature refers to its being an integrated whole of functions and traits and not to brute otherness, that the rationale for keeping functions and traits off-limits to our intentional action involves illicit power over others, and that the nature that is to be kept off-limits is, at least in part, socially constructed and the product of human activity.

Creation, Eschatology, and Biotechnology

From the standpoint of Christian ethics, it seems almost inevitable that that the question of the normative status of human nature in the context of biotechnology will be adjudicated in terms of creation and eschatology. In view of the implications of biotechnology for human nature, the question whether biotechnology in principle violates the nature God created, is at least in principle capable of respecting it, or in some way furthers the divine work of creation is unavoidable, as is the question whether it potentially plays a role in bringing creation to its eschatological destiny, proleptically anticipates aspects of its destiny, or offers only false semblances of it. As they are formulated by Christian thinkers, the first (this chapter) and third (Chapter 4) of the four versions of the claim that normative status attaches to human nature that this book examines present distinct and incompatible answers to these questions.

In the case of NS1, the claim that human nature should be off-limits to intentional alteration, selection, and control finds support in one classical Christian understanding of creation. If creation is thought to be both good, as Genesis 1:31a proclaims ("God saw everything that he had made, and indeed, it was very good"), and completed (that is, a finished work), as one might infer from Genesis 2:1 ("Thus the heavens and the earth were finished, and all their multitude"), then it seems plausible from a Christian theological perspective to suppose that nature, including human biological nature, is something we have not made or imagined, and that respect for God as Creator requires us to leave it as it is. From this perspective, to set out to remake human nature would be to question its goodness and to deny its status as a finished work of God. This argument from Scripture can be supported and rendered more determinate by articulating it in the terms of what we may describe as a broadly Augustinian approach to creation.[15] This approach has four features that are relevant to the issue at hand: (1) creation is identified with *creatio ex nihilo*, understood as the original divine act of creation out of nothing, which, at least ideally, is to be thought of as an instantaneous act, accomplished once and for all, and not an act that extends over a temporal sequence; (2) the act of creation is carried out by God alone apart from the creaturely agencies God brings into being; (3) *creatio continua*, the ongoing work of creation, is identified with God's providential preservation and governance of the finished creation, and not as a work that continues or completes an unfinished work of creation; and (4) the eschatological transformation of creation that is the latter's ultimate destiny occurs at the end of history and is not approximated, much less realized, in history or through historical processes (including, of course, the historical development of technology).[16] These four features make intelligible, and may even require, the claims that nature, including human biological nature, is that which we have neither made nor imagined, but which simply confronts us, and that whatever we do in biotechnology (or any other endeavor, for that matter), we should leave nature, including our biological nature, as it is.

This broadly Augustinian version of NS1 preserves some important classically Christian convictions, but the first two of its four features seem to exempt human nature from change and thus to invite the fourth initial objection to NS1 (namely, that human nature is in part the product of human activity). But the exemption of creation from temporal becoming and the influence of creaturely activity could be thought to apply not to created things but rather to the created order in which these things exist. On this view, created things are neither static nor inert, but their changes

---

[15] By "broadly Augustinian" I have in mind something like an ideal type consisting of certain Augustinian convictions rather than a comprehensive or precise statement of Augustine's actual doctrine of creation.

[16] A propos of the first two features, Augustine's struggle to account for the extension of God's creative act as depicted by the six-day schema of Genesis 1 betrays his convictions that the creative act of an eternal and omnipotent God is most properly thought of as instantaneous and as executed in full without employing creaturely agency. See especially Augustine, *City of God*, 11.30 and idem, *On the Literal Interpretation of Genesis*, 3.7. Of course, Augustine's position is a complex one, and it is disputable how central or basic these convictions are to it.

and activities are intelligible only in terms of a created order that is finished and unchanging. In that case, the assumption that human nature is given once for all in an initial creative act is avoided while anthropogenic changes to human nature can be accommodated.[17] O'Donovan's version of NS1 formulates the Augustinian position along these lines. Although, as we will see, O'Donovan's discussions of human biotechnology are inconclusive, his version of NS1 deserves our attention because it is the most prominent Augustinian approach to human biotechnology and because it has influenced the stances toward biotechnology of other important Christian ethicists whose positions are more conclusive.[18]

Creation and Eschatology

At the center of O'Donovan's moral theology is the claim that Christ's resurrection both vindicates the moral order that is inherent in creation, securing its moral status in the face of the threat posed to it by the fall, and reveals the eschatological transformation that is the destiny of creation and the ultimate meaning of history. I will take these two points in order, beginning with created order and then turning to the eschatological transformation of creation.

---

[17] This understanding of NS1 finds partial support in Augustine's own position. For Augustine, the kinds God has created do not rigidly determine the particular characteristics of the individual things that exemplify those kinds but rather delimit a range of possible characteristics, some of which are then actualized by God in accordance with the divine will and wisdom, as God brings individuals of that kind into existence. See Augustine, *On the Literal Interpretation of Genesis*, 6.13.23–6.18.29. This indeterminacy of kinds explains why individual humans, while sharing the same kind, are not identical to each other but exhibit a variety of characteristics. Each individual manifests God's actualization of some of the particular characteristics that are consistent with the human kind. This account of diversity within kinds seems capable in principle of accounting for unintentional anthropogenic changes to human nature and even allowing for intentional changes while retaining the conception of human nature as "a world which we have not made or imagined, but which simply confronts us" insofar as (1) the fact that the unintended changes to human nature that we have pointed out (namely, those involving height, longevity, and brain functioning) do not amount to a transformation of human beings into some other kind but rather exhibit part of the variety of characteristics that are consistent with being human, and (2) the likelihood that this will also be the case with the intentional changes brought about by biotechnology in the foreseeable future. However, Augustine does not attribute any role to creaturely activity, whether intentional or not, in actualizing the possibilities that are inherent in kinds. It is God alone who actualizes them. But without a role for creaturely activity, the Augustinian position cannot account for the unintended changes within the human kind that we have referred to, much less justify intended changes. Thus, Augustine's own position does not avoid the fourth initial objection to NS1, and it does not authorize intentional changes to human biological nature.

[18] See Robert Song, *Human Genetics: Fabricating the Future* (Cleveland: Pilgrim Press, 2002); Brent Waters, *From Human to Posthuman: Christian Theology and Technology in a Postmodern World* (Aldershot, UK: Ashgate, 2006); and idem, *Christian Moral Theology in the Emerging Technoculture: From Posthuman Back to Human* (Aldershot, UK: Ashgate, 2014).

O'Donovan understands creation as a finished work that creaturely action presupposes and does not contribute to. "That which most distinguishes the concept of creation is that it is complete... 'Created order' is that which is not negotiable within the course of history, that which neither the terrors of chance nor the ingenuity of art can overthrow."[19] Creation is complete in the sense that it is an ordered whole. Specifically, O'Donovan argues that created things exist in *generic* and *teleological* relations – meaning, respectively, that they belong to *kinds* (as, for example, individual human beings belong to the human kind) and are ordered to *ends* (as, for example, vegetables according to Genesis 1:29f. are ordered to humans and other animals for food and humans are in turn ordered to God as their ultimate end). However, creation is also complete in the sense that it is a finished act of God. The point is not that creation is static or that created things came about through an instantaneous divine act. Rather, it is that (1) whatever processes God might have used to bring created things into existence, these things constitute an ordered whole that consists of generic and teleological relations; (2) this ordered whole itself is unchanging and thus finished; and (3) it is this order that explains what has come about in time, and not vice versa: "Creation as a completed design is presupposed by *any* movement in time."[20] That any things have come into existence through temporal processes and that they exist in ordered relations is explained by the ordered whole God created and not by those processes, which presuppose that order rather than generate or explain it. The things that are ordered are temporal and changing; they do not share the unchanging and finished character of the generic and teleological orders in which they exist. O'Donovan accepts that humans and vegetables came into existence through temporal processes and may eventually pass out of existence, but what they are – their nature – is constituted by the generic and teleological orders in which they exist, which account for what they are. These claims (which are somewhat underdeveloped but nevertheless plausible) clarify the senses in which O'Donovan affirms the first Augustinian feature and in which nature for him is "a world which we have not made or imagined, but which simply confronts us." What confronts us in nature, including human nature, is not an indeterminate or brute givenness that stands over against our action but a completed generic and teleological order. And what biotechnology (like every human

---

[19] O'Donovan, *Resurrection and Moral Order*, pp. 60f. See also O'Donovan, *Begotten or Made?*, pp. 12, 28–30.
[20] O'Donovan, *Resurrection and Moral Order*, p. 63.

action) must respect is not the brute givenness of things (and thus not any allegedly pure or original state of things) but the orders in which they exist.

O'Donovan thus denies that created order is negotiable within history, but he stresses that it is destined for an eschatological transformation. As we will see in Chapter 4, the prospect of the eschatological transformation of nature is sometimes invoked to support technological transformations of nature in the present, under the assumption that these transformations can be plausibly understood as approximations to or partial realizations of the final transformation. O'Donovan rejects this move. He insists that the eschatological transformation of creation occurs at the end of history and is not progressively realized or approximated in history or through historical processes (and thus not through biotechnology). O'Donovan does think that the eschatological destiny of creation is visible in history in the form of institutionalized practices that proleptically anticipate the eschatological transformation of creaturely orders.[21] But to anticipate the ultimate destiny of creation is not to transform creation in accordance with its ultimate destiny. History for O'Donovan manifests God's providential vindication of the order God has created and certain anticipations of its eschatological destiny but not any progressive approximation to that ultimate destiny.

Created Order and Biotechnology

O'Donovan's affirmation of the created order as a finished work combined with his denial that its eschatological destiny is approximated in history rule out a positive Christian theological justification for intentional interventions into human nature of the kind we will encounter in Chapter 4. But his acknowledgment that an unchanging created order is compatible with variation and change among the created things that are so ordered appears to accommodate some possibilities that he does not explicitly

---

[21] His clearest example points to marriage as an embodiment of created order and celibacy as the anticipation of its eschatological destiny, in which "they will neither marry nor be given in marriage." See O'Donovan, *Resurrection and Moral Order*, pp. 71f. But O'Donovan stresses that each of these states (marriage and celibacy) has its distinct institutional form and retains its own integrity, so that celibacy does not exercise any transforming pull on marriage this side of the eschaton. Marriage embodies created order while celibacy *anticipates* the eschatological destiny of marriage but does not transform it such that celibacy now *approximates* the eschatological destiny of marriage. In general, O'Donovan stresses that the generic and teleological relations that constitute created order retain their own constancy in history; they are neither gradually nor suddenly transformed in accordance with their eschatological destiny, though (as in the case of marriage and celibacy) they are relativized with respect to that destiny by the presence alongside them of institutionalized anticipations of their final transformation.

consider. First, it clearly allows him to account for the unintended changes to human nature, including the examples of literacy, height, and longevity that we identified previously. Does it also allow for intentional alterations of functions and traits, so long as these alterations respect the generic and teleological relations within which humans exist and do not presume to approximate the eschatological destiny of human nature (whatever it would mean in practice to meet these conditions)? If created order is compatible with variation and change among created things, then intentional alterations of created things cannot be prohibited on the grounds that they violate a finished creation. Might they then be permitted on the grounds that they contribute in some way to created order?

O'Donovan does not draw any such conclusion. But a positive answer to this question might take the following form. First, because O'Donovan's notion of created order leaves room for temporal causal processes by which actual created things come into existence and take on their characteristics, he can account for the role of unintentional human actions in shaping human nature as it now is. Second, O'Donovan's conception of created order appears to allow for intentional human alteration of nature on a limited scale. O'Donovan points out how natural science abstracts from the primary generic and teleological orders in which things exist to concentrate on secondary orders. For example, scientific method ignores the obvious (and normatively significant) ordering of vegetables to human and nonhuman animals for food to focus on previously unnoticed (and normatively insignificant) generic orderings, such as the one in which animals and vegetables alike are members of the class of things that are subject to genetic and evolutionary processes.[22] O'Donovan does not explicitly claim that by attending to these less obvious and normatively insignificant generic orderings, science and technology may also discover new things about already-known, morally significant generic and teleological relations. But that claim seems to be not only consistent with his position but consonant with what much scientific research does. By ignoring the obvious teleological ordering of vegetables to humans for food to concentrate on their nonobvious generic ordering as things subject to molecular processes, scientists may (and often do) discover previously unknown teleological orderings of vegetables to humans for, say, medicine or biofuel. It is true that this knowledge is attained by methodologically abstracting from the obvious generic and teleological orders in which vegetables stand and treating them as if they were ordered only by nonobvious relations such as those of

[22] O'Donovan, *Resurrection and Moral Order*, pp. 48–50.

genetics and evolution. But (at least in this case) the abstraction is ultimately undertaken for the sake of the obvious orders. Abstracting from the obvious teleological ordering of vegetables to humans for food, biotechnology concentrates on the nonobvious generic ordering of things subject to molecular processes, resulting in the discovery of new dimensions of the obvious ordering of vegetables to humans. From here, it is only a small step to the claim that the resulting knowledge of vegetables facilitates technological intervention into their biological properties to render them suitable to use as medicines or biofuels. In sum, biotechnology seems capable on O'Donovan's own account of respecting and even contributing to the teleological ordering of vegetables to humans that is part of God's finished work of creation.

If this argument, which O'Donovan does not himself make, is consistent with his position, might it not also justify some alterations of human biological functions or traits? We can at least imagine changes to functions or traits that would not violate the generic relation of humans to one another or their teleological relation to God, while they might well enhance aspects of these relations. For example, an alteration of emotion that renders humans more sociable might enhance generic relations of humans to one another. If so, it would be possible at least in principle to intentionally alter human biological functions and traits in ways that do not violate, but rather serve, the generic and teleological orders in which humans exist. However, there are two reasons why O'Donovan would probably not endorse the line of argument I have just sketched. Both reasons involve biotechnology as a certain kind of human action, namely, one that treats the object of intervention as material to be made. The first reason issues a broad warning to all scientific and technological approaches to nature. The second reason, which is elaborated in the next subsection, focuses on the specific wrongness of interventions involving humans.

The first reason does not rule out scientific and technological approaches to nature, but it issues a strong warning along a familiar line of twentieth-century criticism of science and technology.[23] While O'Donovan accepts the legitimacy in principle of scientific abstraction from known orderings to attend to previously unknown orderings such as those of living things

---

[23] O'Donovan, *Begotten or Made?*, p. 3. While O'Donovan's critique is consistent with the Augustinian themes detailed previously, it is, as he notes, deeply indebted to Jacques Ellul and George Grant. Grant in turn was strongly influenced by Heidegger's understanding of technology as the disclosure of being as "standing reserve," that is, as material at the disposal of humans to order at will (in which they too incur the danger of being taken as standing reserve). See George Grant, "Thinking about Technology," in *Technology and Justice* (Concord, ON: House of Asansi Press, 1986), pp. 11–34; and Martin Heidegger, "The Question Concerning Technology," in *The Question Concerning Technology*

subject to genetic and evolutionary processes, he warns against presuming that the kinds and ends discovered by science are the only kinds and ends there are. While the procedures that focus on the generic ordering of all living things as subject to genetic and evolutionary processes or to more basic chemical and physical processes are legitimate in their own place, they lend themselves to the assumption that creation is no more than undifferentiated mass and energy or mere matter available for willful imposition of form. O'Donovan of course opposes this assumption. "We must understand creation not merely as the raw material out of which the world as we know it is composed, but as the order and coherence *in which* it is composed,"[24] and the fundamental decision faced by the human agent is the decision between realism and nominalism: whether to regard this order and coherence as real or as merely imposed by the human will (and thus available to be reordered at will).[25] In short, while scientific abstraction from known orderings is in principle legitimate, O'Donovan worries that it is all too easily taken to imply that order is merely what scientific ingenuity imposes on the world.

As with science, so with technology. For O'Donovan, the sharp distinction between creation as a real order and as raw material available to the human will-to-form corresponds to a broadly Aristotelian distinction between two kinds of human action, namely, " 'acting" properly understood, which responds to generic and teleological orders as created by God and respects them as such, and "making," which treats created things as unformed matter available for human fashioning. Making, as O'Donovan understands it, characterizes technological action in a broad sense, of which biotechnological determination of human nature is one expression. The broad sense indicates that what is most fundamentally at issue is neither technological devices nor the discrete acts in which they are employed but rather the mind-set in which a culture "thinks of everything it does as a form of instrumental making." As a mind-set or stance toward reality, technology approaches the world not as a completed order but as material for human fashioning. "When every activity is understood as making, then every situation into which we act is seen as a raw material, waiting to have

---

and Other Essays, translated and with an introduction by William Lovitt (New York: Harper and Row, 1977), pp. 3–35. Heidegger understood technology in terms of the late industrial era; his paradigm was the hydroelectric plant, which gathers and stores energy, making it available for distribution. For O'Donovan, technology is fundamentally nominalistic; it presupposes that nature has no intrinsic order but is simply waiting, as it were, for whatever order humans will to impose on it.

24 O'Donovan, *Resurrection and Moral Order*, p. 31.

25 Ibid., p. 35.

something made of it."[26] To approach nature in this way, as O'Donovan thinks modern societies do, is to treat it as formless and void, awaiting our creative act, in which we substitute our own will for the divine will and thereby act as quasicreators who do what God the Creator alone properly does and has done.

These warnings indicate that biotechnological interventions may fail to respect created order not by directly violating it but by virtue of the kind of action they involve (namely, making). They therefore suggest a version of NS1 that accommodates unintended changes to human nature while casting doubt on intentional changes. Yet they are warnings, not prohibitions. O'Donovan acknowledges the legitimacy of scientific procedures and appears to allow for some technological remaking of nature. It is only a culture that "thinks of *everything* it does" as making or an understanding of "*every* activity" as making that is problematic. These qualifications suggest that intentional alteration of nature is not in principle incompatible with respect for created order. The danger is not in scientific and technological approaches to nature as such, but in the tendency to default to these approaches, when it is a matter of determining the nature and significance of created order and the proper stance of human agents toward it.

In sum, the first reason why O'Donovan might not accept a line of thought that combines biotechnology with created order is inconclusive. It raises doubts about the compatibility of biotechnological interventions with respect for created order but does not prohibit such interventions. The second reason, however, suggests that biotechnological determination of human characteristics is incompatible with respect for created order in a way that determination of the characteristics of other creatures is not. Some things, we might suppose, *are* teleologically ordered to humans as things to be made by them. Perhaps O'Donovan would, along the lines suggested previously, concede that vegetables are ordered to humans in this way. However, it would not follow that the determination of *human* biological nature can be justified in these terms. Each human being is related to every other human being generically (as members of the human kind) but not teleologically (no human exists to serve another human's purposes). No human being is ordered to another human being as something to be made (or remade) by the latter. If intentional determination of biological functions and traits counts as making, then O'Donovan appears to rule it out as incompatible with the generic ordering of humans to one another.

---

[26] O'Donovan, *Begotten or Made?*, p. 3.

Begetting and Making Children

This point brings us to O'Donovan's influential distinction between treating children as "begotten" and as "made" in the act of procreation. The distinction of course recalls the analogy the Nicene Creed draws from the creaturely world in declaring the Son to be "begotten not made." Just as that which we beget is (generically) like ourselves, sharing our nature and our status, while that which we make is (teleologically) subject to our purposes and at our disposal, so the eternal Son of God is of the Father's being, not of the Father's will. O'Donovan notes the limitations of the analogy: Begotten humans are not of one being with their parents as the Son is with the Father; begotten humans are also made by God, as the Son is not; and so on. But his point in referring to the analogy is to indicate what he thinks occurs when we approach procreation technologically, as a matter of making rather than acting. As we will soon see, O'Donovan is vague as to what counts as making in the context of procreation (he seems more concerned with an attitude or stance to procreation than with any particular technology). But he is clear that to the extent that we approach procreation as making, we assert ourselves as masters over those whom we generate rather than treating them as our equals and sharers in our nature. "A being who is the 'maker' of any other being is alienated from that which he has made, transcending it by his will and acting as the law of its being."[27] To make rather than beget a child is thus a serious wrong. It fails to treat the child who is generated in accordance with the generic and teleological relations in which she exists, regarding her as a thing rather than a fellow human person and as ordered to other humans rather than to God alone as her end. It thus violates the creaturely being of the child, and a wrong of this kind cannot be made right by the benevolent motives of the parents or by any alleged benefits that might accrue to the child.

It is important to stress that O'Donovan is not claiming that a child who has been generated in such a way is an artifact and not truly a human person. He would not, for example, worry that a child whose genotype is the result of design or selection by others would be something other than a human being just as those who chose it are. To violate someone's creaturely being is not to destroy it. Indeed, the wrong done to the child consists precisely in failing to treat her in a way that respects what she is (or will be, in the case of interventions carried out on gametes), not in making her something other than she is. These qualifications, however, indicate the sense

---

[27] Ibid., p.2.

in which, for O'Donovan, technological determination of human nature involves the illicit power of some human beings (in this case, parents) over other human beings (in this case, their children).

If this is the sense in which "man's power over nature is the power of some men over other men with nature as their instrument," which acts does O'Donovan count as acts of making rather than begetting? O'Donovan introduces the distinction between begetting and making in relation to reproductive technologies generally, saying little about biotechnological enhancement more narrowly. However, what he initially says about reproductive technologies is comprehensive, targeting not particular technologies but technological assistance in reproduction as such. Elaborating the distinction between begetting and making, O'Donovan contrasts "acting together," in which a procreating couple generates new life out of their sexual union and entrusts the results to divine providence, and "fashioning the future," in which a couple takes measures to control the processes of procreation to ensure to the extent possible the most favorable outcome.[28] If meant to be decisive, this contrast would almost certainly include parental determination of their child's biological characteristics in the category of the reproductive technologies that he seems to rule out. Moreover, O'Donovan clearly considers interventions to be justifiable only on the condition that they aim at curing or compensating for pathologies.[29] Therapeutic interventions (presumably) do not make or remake nature but merely anticipate or respond to threats to it. It seems likely, then, that O'Donovan would count intentional alterations of nonpathological human functions and traits (that is, enhancement) as instances of making rather than acting and thus as violations of the generic order of humans to one another.[30]

---

[28] O'Donovan, *Begotten or Made?*, pp. 7f. See also ibid., p. 17. As this point indicates, it is not only God's role as Creator that is usurped by technological action. The attempt to control the effects of our actions also takes the reigns of providence, which disposes over the order of creation in its historical actuality.

[29] See especially O'Donovan, *Begotten or Made?*, pp. 4, 6, 31f., 68, 70f. O'Donovan sometimes justifies this restriction by appealing to the normative status of human nature as the finished work of God that intentional actions in the domain of biomedicine, as elsewhere, must respect as it is given. (See O'Donovan, *Begotten or Made?*, pp. 5, 12, 16, 19.) Here, he reverts to a conception of human nature as that which is merely given. A more plausible justification would appeal instead to the threat to creation in its fallen state. Uses of technology to protect or restore biological malfunctions or even to compensate for them are aimed at protecting or restoring human nature in the face of threats to it. At least in principle, these uses differ from making.

[30] It is likely, but not certain because O'Donovan permits some reproductive technologies, including in vitro fertilization for married couples, even though his contrast between "acting together" and "fashioning the future" appeared to rule it out from the outset. If that contrast is not decisive in the

However, the matter is not so simple. On O'Donovan's own account the created order in which parents and children interact seems to be more complex than his simple distinction between begetting and making indicates. O'Donovan describes created order as "pluriform," which means that in any situation of choice created order is present in a multiplicity of generic and teleological relations that comprises the "moral field" in which an agent acts.[31] Right moral action requires the agent to determine which of these multiple relations takes priority in a situation of choice. O'Donovan does not explicitly discuss the moral field in which determination of the biological traits of children is located. However, we can easily identify several generic and teleological relations that are present in this field in addition to the generic relation in which parents and child exist as fellow human beings. Although it is true that, qua human, parents and children are related to one another generically but not teleologically, it is also true that "parent" and "child" are generic orders in their own right, and it seems clear that there are complex teleological relations between these two orders. For example, children are under the authority of their parents, but parenthood is for the sake of the child's good, and so on. If these relations are indeed present in a situation of moral choice, then right action on O'Donovan's own account would require discerning how these

---

context of reproductive technologies, it may not be decisive in the context of enhancement technologies either.

[31] Created order is "pluriform" in the sense that particulars exist in multiple generic and teleological relations with their characteristic goods. These relations and their goods, while fixed, will be actualized in diverse ways in accordance with concrete circumstances, which will require that some relations and their goods take priority over others in the moral choices agents must make in these circumstances. Thus, O'Donovan distinguishes between "created order," which is unchanging, and the "moral field" as the concrete context in which the agent acts (See O'Donovan, *Resurrection and Moral Order*, p. 191). To illustrate this point, he refers to a hypothetical case in which one is faced with the question of whether to intervene in the matter of a friend who appears to be engaged in an extramarital relationship. In such a case one must determine the requirements of the various generic relations in which one exists (e.g., with one's friend and with the friend's spouse) and the priorities among them (e.g., whether the loyalty required by one's long-term friendship with the friend takes priority over that required by one's more recent friendship with the friend's spouse, or whether their marriage now requires the same loyalty to both, and so on) as well as the relevant goods (critical judgment or unjudging support of the friend, truth to the spouse, and so on). The relations and their goods (created order) are constant in the sense that both the basic features of friendship and the goods inherent in it (e.g., loyalty, unjudging support, critical judgment) are unchanging, but which of these features or goods should be actualized in one's present action will differ in accordance with circumstances (moral field). History for O'Donovan therefore manifests a diversity of actual orderings in which some of the generic and teleological relations in which a particular exists and their characteristic goods will be concretely actualized in the agent's moral choices while others will not. This relationship between created order and moral field enables O'Donovan to account for the dynamism and diversity of the historical circumstances of human action while maintaining the unchanging nature of the kinds and the moral requirements they imply.

different relations are ordered in that situation. Would an act in which the parents attempt to form their child count as a violation of her generic identity with them as their fellow human being (in which she exists as her parents' equal and does not exist for their purpose) or, alternatively, would such an act count as an instance of the teleological ordering of parenthood to the child's good? This question arises frequently in the context of parental intervention into their growing child's environment. Why, then, should it not also arise in the context of intervention into the child's biological nature? May we not assume that the moral field in both contexts consists of the same complex of generic and teleological relations? If so, then why should we assume that O'Donovan would count only the generic relation of parents and children as fellow human beings in the context of intervention into the child's biological nature and not also their generic and teleological relations as parents and children and the responsibility, which is inherent in the latter relation, of parents to secure and promote their children's good?

O'Donovan might respond by arguing that alleged requirements of teleological relations between parents and children cannot justifiably overrule the requirements of the generic relation of parents and children as fellow humans, and that this generic relation is inconsistent with any acts that involve parents making children. But at this point we may ask whether acting and making are as sharply distinct from one another as O'Donovan supposes. Once again, the intervention of parents into their child's environment illustrates the issue. Are the interventions involved in child rearing instances of acting or making? If they are done right, we might be inclined to say, they are both. They are clearly attempts at "fashioning the future" (and thus count as "making"). But they do not treat the child's nature as mere raw material to be shaped by them as they will (and should thus count as "acting"). In sum, these environmental interventions are ambiguous on O'Donovan's terms. They involve both acting and making. But would not the same ambiguity attend at least some acts of determining a child's biological characteristics?

These questions indicate points at which O'Donovan's position (either as he articulates it or in a form that corrects what seems to be an exaggerated distinction between acting and making) may provide openings for biotechnological enhancement. O'Donovan might well block those openings by insisting that medicine should be restricted to the treatment of pathologies (that is, to therapy as opposed to enhancement) or by denying that parents could in fact secure or promote the good of children by altering their biological nature. However, it is clear that his distinction between begetting and making, when understood in the broader context of his conception of created order, cannot by itself justify the claim that to alter or control a child's biological nature is in principle to exercise illicit power over her or to otherwise wrong her. His position, at least as he has formulated it, leaves open the possibility that by enhancing their child's biological characteristics parents could act rightly in the pluriform generic and teleological orders that comprise their moral field and could combine "acting together" with "fashioning the future" in a way that does justice to the confluence of generic and teleological orders that define parenthood.

We may conclude, then, that neither of the two reasons O'Donovan might urge against the line of thought that suggests that biotechnological determination of human nature is in principle compatible with respect for created order can definitely rule out such alteration on those grounds. This conclusion is a significant one for debates over biotechnological enhancement, as it makes clear that what is undoubtedly the most compelling and most influential Christian version of NS1 fails to render human nature off-limits to biotechnological alteration, selection, or control.

Autonomy, Equality, and Genetic Choice

I have now examined the first of three arguments that determination of the biological nature of others is wrong because, as C. S. Lewis put it, it involves illicit power of some humans over others. The illicit power Jürgen Habermas's version of NS1 has in view differs from that of O'Donovan. Habermas aims to show how the selection or design of the genetic characteristics of children by parents does not treat them as things to be made but rather violates their autonomy and equality, thereby undermining two essential conditions for their full participation in public life.

Habermas's specific arguments are inseparable from his political philosophy and from the circumstances in which he formulated them. However, the claim that biotechnological enhancement potentially violates autonomy and equality is widely asserted, and it is not dependent on the particular contexts of Habermas's account. The concern that the one whose genotype has been chosen by another may not be able to understand herself, or be recognized, as autonomous and equal to others, has relevance for any moral or political position for which autonomy and equality matter, while the prevalence of PGD, in which embryos are generated by in vitro fertilization, tested for genetic characteristics, and then typically selected or rejected for implantation on the basis of those characteristics, makes Habermas's concerns relevant beyond the particular debates that prompted them. The following analysis therefore looks beyond the particular contexts of Habermas's arguments and focuses on their broader relevance. Nevertheless, for the sake of clarity, I begin my discussion of his position by briefly describing the circumstances that prompted those arguments and the role his political philosophy plays in his formulations of them. Following the description of these contexts, I identify the crux of Habermas's concern, which is easily misunderstood in ways that invite superficial dismissals of his position and prevent an adequate consideration of his concerns regarding the threat of biotechnology to autonomy and equality. Finally, I examine in succession his arguments regarding the threat of biotechnology to autonomy and equality.

The Context of Habermas's Position

The circumstances in which Habermas formulated his position on the selection and design of genetic characteristics involve debates in Germany in 2000 and 2001 regarding whether legal protections of human embryos enacted during the previous decade should be relaxed to accommodate PGD and research on embryos.[32] However, his argument focuses not on the moral status of the human embryo but on the choice of genetic characteristics of children by their parents by means of PGD. Although he thus dwells on choices made at the embryonic stage, his arguments are relevant to genetic decisions made by parents at any point prior to their child's capacity to consent to such decisions. And although he is aware that PGD is currently used almost entirely negatively, that is, to identify genetic predispositions to diseases, he assumes (not unreasonably) that it will eventually allow positive selection of predispositions to desired traits or capacities, and he attempts to develop an appropriate ethical framework in advance of that capability.

Turning now to his political philosophy, Habermas begins with the liberal assumption that public morality in modern pluralistic societies cannot be grounded in religious or philosophical conceptions of a good human life, all of which lack the universality that is required of publicly binding norms, but must instead be grounded in norms of justice that reflect the interests of all, whatever their conceptions of a good human life may be. In Habermas's version of this liberal assumption, justified norms

---

[32] A decade after Habermas wrote, PGD became legally permissible in Germany in cases in which the likelihood of transmission of a genetic disease or of miscarriage or stillbirth is high; a policy Habermas would presumably endorse.

are arrived at through a communicative process among persons who understand themselves and one another as equal, autonomous subjects. However, the mutual recognition of persons as persons cannot simply be taken for granted. To see ourselves and others as equal and autonomous presupposes that we see ourselves and others as beings that possess the requisite qualities for equality and autonomy. Thus, while public moral norms are neither derived from nor justified by substantive conceptions of a good life, they do depend, in a less direct sense, on a certain anthropological self-understanding among those who consider themselves and others to be bound by them. It is here that Habermas's concern with the ability to design or select the genetic characteristics of others emerges. What happens to this conception of ourselves as equal and autonomous beings when we become aware that our biological nature can be disposed over by others? "This perspective inevitably gives rise to the question of whether the instrumentalization of human nature changes the ethical self-understanding of the species in such a way that we may no longer see ourselves as ethically free and morally equal beings guided by norms and reasons."[33]

The Crux of Habermas's Position

We are now prepared to understand the crux of Habermas's position, which can be expressed in terms of three claims. First, modern political orders depend on the mutual recognition of citizens as autonomous and equal subjects who speak in their own person in the intersubjective determination of public norms. In Habermas's words, "If we see ourselves as moral persons, we intuitively assume that ... we act and judge *in propria persona* – that it is our own voice speaking and no other."[34] Personhood entails the capacity to speak and act in one's own person, and politics depends on the mutual recognition of persons, so understood. Second, drawing on Helmuth Plessner's distinction between being a body and having a body, Habermas argues that "one's own person" includes one's physical embodiment, and thus one's biological constitution. Normative status therefore attaches to one's biological nature as a constituent of one's personhood. Third, Habermas worries that if one's personhood, so understood, is not

---

[33] Habermas, "The Debate on the Ethical Self-Understanding of the Species," pp. 40–41.
[34] Ibid., p. 57. This formulation suggests a notion of moral personhood that is relevant apart from liberal premises. In Christian ethics, to be in a position to act and judge in one's own person can be understood as a condition of one's accountability to God and neighbor.

separable from one's biological nature, then parents who determine their child's genetic characteristics could threaten her capacity to act and speak in her own person and to understand herself as the equal of those who have determined her characteristics.

With these three claims in place, we are prepared to examine Habermas's arguments regarding the threat of biotechnology to our self-understanding as moral and political persons. But those arguments will be clearer to us if we dispense with some common misunderstandings of Habermas's position. First, Habermas does not argue that to choose someone's genetic characteristics is to deprive that person of her autonomy or equality. He is aware that autonomy and equality depend on possessing certain characteristics, not on having acquired those characteristics in one way rather than another. Second, he does not argue (as O'Donovan seems to imply) that one who chooses the genetic characteristics of another directly violates the latter's autonomy or equality. Because he does not believe that one possesses full moral status prior to birth, there is for him no autonomy or equality to be violated in the genetic design or selection of embryos. Rather, his principal argument is that one whose genetic characteristics have been chosen by another may find it difficult to understand herself (and thus to act and judge) as one who is autonomous and equal to others, with the result that her status as a moral and political subject, which presupposes the self-understanding and mutual recognition of persons as autonomous and equal, is threatened.[35]

It is possible to misunderstand Habermas's arguments along a third line. One might suppose that according to Habermas genetic design or selection threatens the capability of a person to think and act *in propria persona* insofar as a person who is genetically designed will understand herself as a kind of artifact rather than a person. At times, Habermas says things in ways that lend support to this view. He stresses that contemporary biotechnological interventions involve a technical mode of action, which experimentally remakes organisms, in contrast to the mode of action involved in therapeutic interventions and classic breeding, which respect "the inherent dynamics of autoregulated nature."[36] As a result, "the categories of what is manufactured and what has come to be by nature ... dedifferentiate" – a process that began with the natural world but now

---

[35] See ibid., pp. 77f. for a clear statement of the difference between a violation of the rights of a person to autonomy and equality, on the one hand, and a threat of harm to the person's self-understanding as one who is autonomous and equal, on the other hand.

[36] Ibid., p. 45.

extends to the world of human action and self-understanding.[37] This concern with the loss of the distinction between what is natural and what is made has much in common with O'Donovan's analysis of technologically assisted reproduction.[38] One might therefore expect Habermas to argue as O'Donovan does that it is wrong in itself to treat a child as an object of manufacture by selecting or designing her genotype. In a brief and tentative foray into theology, Habermas does just this, suggesting that the Christian conception of the relation between the Creator and the creature illustrates the ethical impropriety of humans taking up the stance of makers toward other humans.[39] However, for the most part Habermas denies that to be genetically designed is to be treated as an artifact. And he does not find anything problematic in the remaking of human nature by technical action as such. "My criticism is not rooted in some fundamental mistrust of the analysis and artificial recombination of the components of the human genome as such."[40] It is not subjection to technical reason as such that is the problem for Habermas. A person who subjected *her own* biological nature to technical reason by choosing her own genetic characteristics would not thereby imperil her status as a moral subject.

We are now prepared to understand Habermas's position on the genetic selection and design of children by their parents. According to that position, the threat to our ethical self-understanding occurs when our genetic characteristics (more precisely, our nondisease-related genetic characteristics, which are his concern in what follows) are chosen by *others*; it does not consist in being the object of their technical action as such. Habermas argues that when a child becomes aware that she has been the object of the genetic choices of her parents, she may find it difficult if not impossible to understand herself as autonomous and as equal to those who chose her genetic characteristics. Both our capacity to see ourselves as authors of our own life histories, which for Habermas is central to autonomy, and our existence in symmetrical and reversible relations to others, which for him is central to equality, are threatened when our genetic characteristics are chosen by another person without

---

[37] Ibid., p. 46.
[38] Although it is Weberian, with a focus on technical rationality in the service of subjective values, rather than Heideggerian, with a focus on the way technology orders nature to our use as "standing reserve," Habermas's account of the instrumentalization of our biological nature is similar to O'Donovan's account of the technological reduction of nature to raw material.
[39] Habermas, "Faith and Knowledge," in *The Future of Human Nature*, pp. 114f.
[40] Habermas, "The Debate on the Ethical Self-Understanding of the Species," pp. 86f.

our consent. That is to say, a person whose genetic characteristics have been chosen by another may not be able to see herself as the sole author of her own life history, and thus as autonomous, and she may not be able to see herself as existing in a symmetrical and reversible relation to those who chose her characteristics, and thus as equal to them. Lacking the ability to see herself in this way, she may be unable to take her place as a full participant in moral and political life.

In place of the ontological wrong that for O'Donovan is done to the embryo in its creaturely being, then, Habermas finds a potential psychological harm (with implications for her moral identity) to the conscious person the embryo will, if all goes well, eventually become. Critics have pointed out that this argument is an exceedingly fragile one. It appears to rest on a problematic empirical claim about the future psychological state of the child whose genes have been designed or selected. The problem is that we do not know, and at present we cannot know, how a person who becomes aware that her genotype has been chosen by her parents is likely to see herself, and it is unreasonable to base policies on speculations about unknown factors like these. This criticism is sound, but Habermas's position can be formulated (or reformulated) in a way that avoids it. It is possible to avoid speculations about future psychological states and the problems associated with such speculations by simply asking instead whether a child whose genetic characteristics have been chosen by others *would have a reason* to see herself as unequal or lacking autonomy. In what follows I treat Habermas's claim in this way, namely, as the claim that a child who becomes aware that her genetic characteristics were chosen by her parents would have a plausible reason to question her status as autonomous and equal to others.[41]

---

[41] Like many people who share his worries about the effects of genetic technologies on autonomy and equality, Habermas seems at points to presuppose genetic essentialism (the identification of who a human being is with her genotype, so that to choose a person's genes is to determine her identity) and genetic determinism (the notion that physical or behavioral phenotypes are determined by genes). He also seems to presuppose that the genetic choices made by parents will be irreversible. If these presuppositions are true, then the genetic choices of parents would irrevocably determine the identity and phenotypic characteristics of their children. But in fact, genes play only a partial role in most of the phenotypic characteristics tomorrow's parents are likely to select for or engineer, including physical traits such as height and strength, and performance capabilities involving athletics, music, or cognition. Moreover, depending on how genetic technology develops, children who grow up to resent the genetic choices made by their parents might eventually be able to offset or even reverse those choices. In short, the degree of control that parents exercise in their choices is less than Habermas seems to realize, and the threats these choices pose to autonomy and equality are correspondingly less severe than he suggests. Nevertheless, these considerations do not invalidate his concerns. Even if we take account of these limitations on the parents' choices, the questions regarding autonomy and equality remain. The child whose

Biotechnology and Autonomy

Let us begin with the alleged threat to the child's self-understanding as an autonomous person. The claim that to choose the genetic characteristics of children threatens their sense of autonomy is widely rejected. Bioethicists often point out the oddity of thinking that the choice of a child's geno-type by her parents without her consent would give her reason to question her autonomy. For the alternative to her parents choosing her genotype is not that *she* chooses it but that it results from purely contingent fac-tors. The child is no more in the position of choosing her own genotype (at least not the one she is born with) in the one case than she is in the other. In the acquisition of a genotype there is no autonomy either way. But Habermas's argument is not vulnerable to this criticism. For him, it is only with the awareness that *another person* has chosen one's genetic characteristics that doubt about one's autonomy would arise; by contrast, the sheer contingency of one's origin is in fact a condition of seeing one-self as autonomous.[42] One's understanding of oneself as author of one's life history depends on one's ability to distinguish the root of one's identity in one's "natural fate" as an organism whose origin is not at *anyone's* disposal, on the one hand, from what one has become through one's "socialization fate," which *is* in the hands of others, on the other hand. If others enter directly into one's natural fate through their genetic choices, then one's socialization fate goes all the way down, so to speak; one's natural fate dissolves into it, and there is no ground in which an understanding of one-self as author of one's life history can take root.[43] And without the ability to understand oneself as the sole author of one's life history, Habermas argues, a basic condition of ethical self-understanding is thus threatened. Once again: To see ourselves as moral persons is to assume that "we act and judge *in propria persona* – that it is our own voice speaking and no other."[44]

Here, then, is Habermas's first argument regarding autonomy: The one whose origin is not contingent but has been determined by the genetic choices of others may be unable to understand herself as the author of her life history and thus as autonomous. But is it reasonable to suppose that

---

genes have been chosen will still be significantly affected by those choices, and she is still in some respect the object of the choices of others.

[42] Habermas makes this last point by invoking Hannah Arendt's well-known argument for the importance of natality (that is, the contingency of one's origin as an event no one disposes over) as a condition for consciousness of oneself as an initiator of action.

[43] Habermas, "The Debate on the Ethical Self-Understanding of the Species," pp. 57–60.

[44] Ibid., p. 57.

one can experience one's autonomy as genuine only if it can be shown to have emerged in a space that is entirely devoid of influence exerted by others? Only if one holds to an extreme form of individualism would one suppose so. Would one's awareness that one never inhabited such a space make it impossible or even difficult for one to understand oneself as autonomous? Only if one were to assume that one currently lacked autonomy for want of such a space. In short, Habermas's worry follows from rather extreme and implausible beliefs about autonomy. It is unreasonable to assume that a person whose genetic characteristics had been designed or selected would hold such views and would thus see herself as lacking in autonomy.[45]

Habermas seems, at least at times, to acknowledge this point. He frequently puts forth a second argument, which involves the weaker but more plausible claim that such a person may find it difficult to understand herself as autonomous, not because of her awareness that her parents have chosen part of her biological nature, but because she fails to identify with the particular biological nature that has been selected or designed by them. In such a case, she may be unable to understand herself as the author of her life history not for lack of an origin that is contingent and exempt from socialization (as the first argument held) but because she happens not to endorse the particular traits that have been chosen for her or the form of life those traits have equipped her for. In her biological nature she is, at least in part, the product of preferences of others that turn out not to be her own preferences, and it is in this sense that she feels she is not fundamentally the author of her own life. By contrast, a person whose genetic characteristics have been chosen by her parents but who, upon becoming aware of their origin, endorses their choices, would presumably not question her autonomy. Her second-order preferences would endorse the first-order preferences that are embodied in her characteristics, and she would therefore act and judge entirely *in propria persona*.

How reasonable is it to worry that the child will fail to identify with the genetic choices made by her parents? (Recall that we are not asking about the empirical probability, which seems low, that a child will fail to endorse

---

[45] In his criticism of Habermas, Robert Song suggests an alternative conception of autonomy that rejects "the fantasy of self-authorship" in favor of "negotiation and acceptance of one's identity, selectively appropriating the things that can be changed, and coming to terms with … those aspects that cannot." Robert Song, "Knowing There Is No God, Still We Should Not Play God? Habermas on the Future of Human Nature," *Ecotheology* 11 (2006): 201. Song's conception of autonomy is more reasonably attributed to a person who becomes aware that her genetic characteristics were chosen by others, if only because it is more consonant with the *actual* ethical self-understanding of most people.

those choices; we are asking whether she might have a reason not to endorse them.) Habermas assumes that when parents choose their children's non-disease-related characteristics they are simply imposing their own arbitrary preferences on the child. "In making their choice, the parents were only looking to their own preferences, as if disposing over an object."[46] There is no reason to suppose that the child will come to endorse these preferences. This situation contrasts with one in which a genetic intervention is undertaken to treat or prevent a disease. "As long as medical intervention is guided by the clinical goal of healing a disease or of making provisions for a healthy life, the person carrying out the treatment may assume that he has the consent of the patient preventively treated."[47] In the case of disease and health, the parents' action is not in the service of their merely subjective preferences but rather of a goal (namely, treatment or prevention of a disease) that we are justified in presuming will be endorsed in the future by the person the embryo will become. The future person who the embryo is, is thus treated, at least virtually, as a second person who consents to the change in her biological nature. "The presumption of informed consent transforms egocentric action [that is, action in the service of the parents' subjective preferences] into communicative action."[48] In the case of genetic therapy, then, where what is at stake is the prevention or treatment of diseases, there is no cause (assuming that the therapy succeeds) for worry that the child will in the future fail to endorse the genetic choices made by her parents and will come to doubt her autonomy as a result of that failure.

It should now be clear that for Habermas it is the future consent of the child that determines whether the genetic choices of her parents threaten her ability to understand herself as autonomous.[49] Because future consent can be reasonably presumed in their case, genetic interventions on embryos aimed at treating or preventing diseases pose no risk of a later failure of the grown child to endorse the changes intended by her parents. The reason is that, unlike interventions aimed at positive enhancements,

---

[46] Habermas, "The Debate on the Ethical Self-Understanding of the Species," p. 51.
[47] Ibid., p. 52.
[48] Ibid.
[49] Habermas draws heavily on Helmuth Plessner's account of the priority of being a body to having a body, but if in the end the only ethical obstacle to genetic design and selection is the consent of the one who undergoes it, then Habermas is committed to the liberal thesis that the body is the object of one's choices, and that thesis prioritizes having a body over being a body. There is, to be sure, a profound moral difference between my body as the object of my choice and as the object of someone else's, and this is of course what Habermas wants to establish. But Plessner's insight would seem to require that my body is the object of my choice, and not of someone else's, because I am first of all identical with it in a way that precedes choice.

those aimed at the prevention or treatment of disease do not merely reflect the subjective preferences of the parents. But can Habermas simply assume that the positive genetic preferences of the parents are merely subjective, and not preferences that we may be justified in expecting the child will come to endorse? Is it only in the case of disease-related interventions that consent may be reasonably presumed? Many bioethicists argue that some enhancements, such as increased physical strength and cognitive capacity, may (at least up to a point) be considered "basic" or "all-purpose" goods insofar as they would seem (in contrast, say, to an increased aptitude for music) to be beneficial for any way of life a child might eventually choose. They go on to suggest that parents would be justified in presuming the future consent of their child in designing or selecting for these basic or all-purpose enhancements.[50] Habermas rejects this argument. He denies that we can presume the future consent of the child for *any* positive traits, no matter how generally desirable they may seem, insisting that it is wrong to assume that high intelligence, sharp memory, or even great physical strength will be endorsed by those we equip with them.[51] "Attributing such consent can only be justified in cases in which there is a certain prognosis of extreme suffering. We can only expect a consensus among otherwise highly divergent value orientations in the face of the challenge to prevent extreme evils rejected by everybody."[52]

With this rigorous standard for presumed consent, which can be met only by interventions that prevent or treat serious diseases, Habermas renders all non-disease-related traits off-limits to genetic design or selection by parents. If he is right that positive genetic choices made by parents reflect no more than arbitrarily subjective preferences on their part, then a child might have a reason not to endorse those choices and thus to be hindered in her ability to consider herself as one who acts and judges *in propria persona*. For if preferences are arbitrary in this way, there is no reason to assume that any two people will share them. But this argument requires us to concede that positive goods of our biological nature are in fact no more than merely subjective preferences. That concession is a high price to pay for keeping parents' hands off their children's genes, and it is also highly implausible. Even if one accepts Habermas's view that all such preferences are ultimately subjective, it would not be difficult for advocates

---

[50] See Allen Buchanan, Dan Brock, Norman Daniels, and Dan Wikler, *From Chance to Choice: Genetics and Justice* (New York: Cambridge University Press, 2000), pp. 167–70.

[51] Habermas, "The Debate on the Ethical Self-Understanding of the Species," pp. 85f.

[52] Ibid., p. 91.

of liberal eugenics (that is, those who argue that genetic improvement is permissible so long as it is carried out without coercion), against whom Habermas directs his argument, to identify positive genetic characteristics that nearly everyone would want to possess. Higher cognitive functioning, improved eyesight, and increased muscular strength would, at least in modest amounts, fit this description. In the case of these and other widely desired characteristics it is unclear that those we equip with them would have reason not to endorse them. We would, on Habermas's grounds, be justified in presuming consent for them.[53]

Biotechnology and Equality

In sum, the worry over our self-understanding as autonomous persons appears to be either exaggerated or unfounded. There is no reason to think that a child would be unable to think of herself as autonomous just because her characteristics were chosen by another (Habermas's first argument) or that she has reason not to identify with the characteristics that were chosen (his second argument), so long, at least, as the enhancements involved moderate improvements to basic functions or traits. But what can we say about the worry over equality? Proponents of genetic design and selection by parents often argue that genetic control over children is no different in principle from environmental control. They point out that profound asymmetries are present in the influences parents exert over their children through control of their environment, and if these environmental asymmetries are not thought to be incompatible with equality, why should we think that genetic asymmetries are? Against this view, Habermas argues that genetic choices made by parents establish an asymmetry between parents and their child that – in contrast to the asymmetry of child rearing – is permanent and thus incompatible with their ultimate equality.[54] With these choices, he claims, parents dispose over their child's biological nature and thereby set themselves in an irreversible relation to her. The child can take up a revisionary posture toward her upbringing, which can be resisted and finally rejected as the child matures, in a way that she cannot toward her genotype, which is permanent.

---

[53] If one holds, contrary to Habermas, that some things people prefer are objective goods, the analysis becomes more complicated because we would have to determine whether the general or all-purpose goods biotechnology promises to bring about are genuine goods. But that consideration takes us beyond the argument Habermas is making.

[54] Habermas, "The Debate on the Ethical Self-Understanding of the Species," pp. 63–65.

This argument is questionable in light of the limitations of genetic choice. For one thing, we now know that the contribution of genes to complex traits such as longevity, cognition, and physical strength is a rather limited one, so that environmental interventions may well turn out to allow for more effective and extensive influence on these traits than will ever be possible with genetic interventions. Moreover, to claim, as Habermas does, that even the most intractable environmental influences are reversible in a way that genetic influences are not is questionable on both ends. On the environmental end, the deep-seated psychological effects of one's upbringing are often impossible to undo, while on the genetic end, depending on the progress of research, selections or alterations of genes may turn out to be reversible. Of course, there is reason to expect that genetic intervention combined with environmental intervention will allow for more effective control than either type of intervention alone will allow for. But if equality is incompatible with the permanence of asymmetries between parent and child, then the rearing of children appears to be no less a threat to equality than is choosing their genes.

Thus far, then, the threat posed by genetic design and selection to our self-understanding as equal seems no more serious than the threat it poses to our self-understanding as autonomous. However, there is a more plausible sense in which the inequality of parents and child involved in choosing genes differs from that involved in child rearing. Habermas touches on it when he observes that in the environmental interventions that constitute child rearing the child plays the role of a second person.[55] The formation of the characteristics that in part make her who she is occurs through interpersonal interactions in which she is the addressee of demands made by her parents, that is, a second person, and in that sense she is their equal despite the obvious asymmetry between them. In contrast with parental expectations that are imposed on children through environmental interventions, Habermas writes, "genetically fixed 'demands' cannot, strictly speaking, be responded to. In their role as programmers, the parents are barred from entering the dimension of the life history where they might confront their child as the authors of demands they address to him."[56] In other words, in the choice of genetic characteristics the child is not

[55] Ibid., p. 62.
[56] Ibid., p. 51. As noted previously, design or selection for positive phenotypic traits differs in this regard from genetic intervention to prevent a disease. In the latter case, Habermas argues, it is, due to the legitimate presumption of consent, a matter of "the clinical attitude of the first person toward another person – however virtual – who, some time in the future, may encounter him in the role of a second person" (ibid., p. 52).

addressed as a second person, and thus as a fellow subject, but is the mere object of their intentional action – and that in a way that does not occur when biological processes determine those characteristics apart from intentional intervention.

If one stresses, as Habermas does, the importance of mutual recognition to moral relations, then this difference between treating a child as a fellow subject whom one addresses and as an object on which one acts is obviously an important one.[57] However, the relevance of this distinction to the moral evaluation of genetic design and selection is not as clear as Habermas supposes. He appears to identify genetic design and selection with objectification and environmental interaction with second-person recognition, but these identifications are imperfect. First, some of the control that parents exercise over their child's environment is thorough enough to virtually nullify recognition of her as a subject, while some forms of parental objectification, such as manipulation, depend on recognition of the child as a subject. In these cases, it is questionable whether the child is truly addressed as a second person. Second, these and other forms of environmental objectification are direct, because they are aimed at an existing child, while some instances of genetic selection and design, such as any that might in the future be carried out on gametes, are merely indirect, inasmuch as they are not performed on an existing child. In sum, the distinction between fellow subject and mere object does not rigorously correspond to the distinction between environmental and genetic interventions (or to the distinction between justifiable and unjustifiable interventions).

We may conclude that despite Habermas's worries, genetic design or selection by parents without their children's consent does not give the children reason to question their autonomy or their equality with those who have chosen their genotypes. Habermas therefore fails to show how biotechnological alteration or control of human nature involves the illicit power of some human beings (once again, parents) over other human beings (their children) and gives us no convincing reason why we ought to confer immunity from design or selection on the genotypes of children and keep them off-limits to intervention.

---

[57] Just as the importance of judging and acting *in propria persona* goes beyond liberal notions of autonomy, treatment of a person as another subject goes beyond modern notions of equality. In Christian ethics, it can be understood as an entailment of respect for the dignity of the person, and it is of course one interpretation of the requirement of Kant's categorical imperative that concerns treatment as an end and not a means only.

Parenthood, Unconditional Love, and Genetic Choice

I have now examined two versions of the claim that determination of the biological characteristics of children by their parents involves illicit power over them, and I have argued that neither version supports the conclusion that these characteristics should be kept off-limits to parental attempts to determine them through biotechnology. It is unclear that such attempts treat those on whom they are performed as things to be made (O'Donovan), and it is doubtful that they undermine the ability of those on whom they are performed to understand themselves as autonomous and equal with those who intervene (Habermas). We turn finally to Michael Sandel. In consonance with his communitarian political philosophy, Sandel holds that attempts to choose the genotypes of children compromise certain norms that are constitutive of parenthood and citizenship as basic social institutions and are morally problematic for that reason.

Sandel treats biotechnology in dramatic terms. For him, it represents "a Promethean aspiration to remake nature, including human nature, to serve our purposes and satisfy our desires." The problem with this "drive to mastery," as he calls it, is that "it misses, and may even destroy ... an appreciation of the gifted character of human powers and achievements."[58] The drive to mastery is problematic for Sandel because it fails to accept the "given" character of our biological nature as that which is (and is what it is) apart from our willful activity. This failure is troubling, he thinks, because acceptance of the givenness of our biological nature is a condition for upholding certain norms that are integral to parenthood, such as unconditional love for children, and to citizenship, such as solidarity with

---

[58] Sandel, *The Case against Perfection*, pp. 26f. Sandel formulates the issue in these terms when he is discussing athletic and musical performance early in his essay on biotechnological enhancement. We commonly refer to people with native athletic or musical ability as "gifted," and by stressing "appreciation" of this quality, Sandel is arguing for the primacy of natural talents and capacities over effort and striving in our valuations of athletic and musical performance. In other words, we should value the display of natural talent more highly than we value the fact that the performer struggled to attain excellence. However, Sandel goes on in the essay to discuss the implications of biotechnological enhancement for parenthood and citizenship, and in these contexts, which are the focus of my attention, the language of gift and appreciation may be less appropriate. The point of Sandel's argument regarding parenthood and citizenship is that we should leave natural capabilities as they are rather than seeking to enhance them. But few of these natural capabilities lead us to descript their possessors as "gifted," and to "appreciate them" will in many cases mean simply accepting them as they are. Perhaps, then, "acceptance of what is given" is closer to what Sandel has in mind in his arguments regarding parenthood and citizenship than is "appreciation of giftedness," and that is what I will assume in what follows. See Sandel, *The Case against Perfection*, pp. 45, 93 for support for this claim that appreciation of giftedness comes down to acceptance of that which is not our own doing.

those who are less fortunate. These norms, he argues, would be compromised if we were to treat genetic and other characteristics not as "givens" to be accepted but as material to be shaped.[59]

Sandel's criticism of mastery is frequently treated – by Sandel and his critics – as a claim about motivation.[60] On this account, biotechnological enhancement is problematic because it expresses or inculcates a desire for mastery over human nature (or perhaps over those whom one enhances). That biotechnology is fundamentally Promethean in this sense is of course a widely held view. But when it is understood in this way, as a claim about motivation, this view is subject to two straightforward objections. One objection is that it is a mistake, and perhaps also an injustice, to attribute this motive to those who seek to enhance their capabilities or those of their children.[61] It is fair to presume that if genetic enhancements become safe and effective many parents will pursue them out of a desire to benefit their children and not out of a desire for mastery over them or their nature. The other objection is that even if those who seek to enhance the capabilities of their children are motivated by a desire for mastery, and that motive is a bad one, it does not necessarily follow that it is wrong to enhance because it may be that the goods enhancement brings about justify it despite the bad motive it involves.[62] These are challenging objections, but it is doubtful that they pertain to Sandel's position, at least in its most plausible form. Although his formulations are ambiguous, it is likely that Sandel thinks of mastery first and foremost as an attitude or stance toward nature, and not a motive for acting to intervene into it.[63] If this is so, then he does not assume that people who choose genetic or other biological characteristics of their children are motivated by a desire to attain or assert mastery over them or their biological nature. To the contrary, Sandel is well aware that many people seek biotechnological enhancements for their children because they want to benefit them by giving them a leg up in life. Indeed, this unassailable motive is one reason why it is so difficult to criticize

---

[59] The contrast between mastery of nature and acceptance of its givenness marks Sandel's position as a classic instance of NS1, while the emphasis on goods integral to basic social institutions such as parenthood and citizenship mark it as an instance of his distinctive political philosophy.

[60] In fact, it is exceedingly difficult to know precisely what Sandel thinks on this matter, if he thinks anything precisely at all. On one page of *The Case against Perfection* (p. 46), he refers to mastery as an impulse, a drive, and a disposition. "Impulse" and "drive" recur frequently in the text. In addition, Sandel refers to mastery as a desire (p. 95), a project (p. 97), a resolution (pp. 99f.), and a stance (pp. 9, 83).

[61] Buchanan, *Beyond Humanity?*, pp. 9, 77.

[62] See Frances Kamm, "What Is and Is Not Wrong with Enhancement?," pp. 93–98.

[63] His most programmatic statement refers to an attitude or stance toward nature. See *The Case against Perfection*, p. 9.

enhancement. However, Sandel argues that whatever their motive may be (whether it is the benefit of their children, their own self-aggrandizement, or Promethean mastery), in choosing the characteristics of their children they treat their children's biological nature as that over which they dispose, and that is a problematic attitude or stance to take toward their children's nature even if their motive is permissible or even admirable. If we understand Sandel's position in this way, as I think we should, it is clear that it offers us a third instance of Lewis's claim that to determine someone's nature is to exercise illicit power over that person.

Genetic Choice and Unconditional Parental Love

I have noted that Sandel thinks that the problem with an attitude or stance toward genetic and other biological characteristics as that over which we may dispose is that acceptance of these characteristics as given is a condition of certain norms that are integral to parenthood and citizenship. Critics have addressed the case of citizenship in ways I find persuasive.[64] I will therefore focus exclusively on Sandel's argument regarding parenthood, which I think has not been treated adequately. Sandel argues that the stance of mastery has a corrosive effect on the norm of unconditional love – a norm that is integral to parenthood. His point seems to be that the child whose biological nature is not designed or selected by her parents but is accepted just as it is can be loved simply for who she is and not because she possesses characteristics her parents deem worthy to equip her with. Correlatively, there is a risk that the latter, conditional form of love will prevail where parents choose the biological characteristics of their children. In her criticism of this claim, Frances Kamm denies that choosing the genetic characteristics of children is incompatible with unconditional love for them. She points out that, just as we may be drawn to a person due to the particular characteristics that person possesses but may end up loving the person rather than his or her characteristics (so that, for example, we will not cease to love the person if he or she loses those characteristics or reject him or her for another person who possesses those characteristics more fully), so choosing desired characteristics of a child who does not yet exist is consistent with loving that child whatever his or her characteristics turn out to be. As Kamm puts it, "[O]ne can know that one will *care about* someone just as much whether or not she has certain traits and yet *care to have* someone, perhaps for their own sake, who has rather than lacks those traits."[65]

This point is a sound one: The choice of the biological characteristics of one's child does not preclude unconditional love for that child and may be undertaken out of concern for the child. However, the question is whether parents committed to unconditional love would choose the biological characteristics of their child in the first place. Here, the analogy with love between adults who choose one another as partners is a problematic one. It is appropriate for adults to be initially drawn to each other based on each other's characteristics because mutual attraction, common interests, shared perspectives, and reciprocal commitment are in part constitutive of the kind of love adults have for one another. But these qualities (that is, mutual attraction, etc.) are not constitutive of the love of parents for children. Indeed, one would gravely misunderstand parental love if one supposed that these qualities were constitutive of it.[66] If by choosing the biological characteristics of their children, the love of parents for their children would come to resemble in this way the love of adults for one another, there would indeed be a threat to the unconditional love that is integral to parenthood, and Sandel would be justified in pointing to acceptance of the givenness of biological characteristics as a condition of that good.[67]

If parental choice of the genetic characteristics of children is compatible with unconditional love for those children, then, it is surely not on the ground Kamm proposes. Ironically, Sandel identifies a more plausible ground when he turns to a distinction between accepting love and transforming love made by William F. May. "Accepting love affirms the being of the child, whereas transforming love seeks the well-being of the child."[68] Following May, Sandel acknowledges that parenthood properly embodies

---

[65] Ibid., pp. 112f. (emphasis added to highlight Kamm's distinction between "caring about" and "caring to have").

[66] The closest analogy to Kamm's proposal would be a case of prospective adoptive parents who compile a list of desirable characteristics and set out to find a child who possesses them but who, having adopted the child, come to love her for her own sake and eventually cease measuring her against their list. Such a case is hardly an appropriate model of unconditional parental love, but it is the model Kamm's argument presupposes.

[67] Sandel's argument is not a deontological argument to the effect that it is never right to violate the norm of unconditional love but rather a consequentialist argument to the effect that it would be a bad thing if unconditional love were diminished or lost. It is therefore decisive only if (1) the effect on parental love identified in the hypothetical does indeed come to pass, and (2) in comparison with the present state of affairs the resulting state of affairs would be detrimental to those involved. As many critics have pointed out, Sandel's assumptions about what will result from biotechnological enhancement are highly speculative, and he considers only negative consequences.

[68] Sandel, *The Case against Perfection*, pp. 49f.

both accepting and transforming love. Good parents accept their children as they are but also push them to become what they are not yet. Unconditional love for children does not, therefore, require that parents merely accept them as they are; indeed, to do that would amount to a failure of parental love, to the extent that the latter also involves transforming love, rather than an expression of it. Given his interest in ruling out genetic design and selection, one might expect Sandel to make a simple allocation at this point, placing intervention into the biological nature of children under their parents' accepting love and intervention into their environment under their parents' transforming love. In accordance with this allocation, parents would accept their children's genotypes as they are given, but would go on to transform what is given by shaping their children's environment. However, Sandel rejects this solution, and rightly so. He is aware that accepting love is needed also in the environmental domain, where parents must both respect and push against their children's limits in developing their capabilities. It is especially important to emphasize this point nowadays, Sandel thinks, when parents are prone to managing every aspect of their children's lives to ensure their success and happiness – a tendency he calls "hyperparenting."[69] In these circumstances transforming love threatens to crowd out accepting love, and Sandel suggests that the parental mentality that is at work in this trend is the same one that fuels interest in the genetic design and selection of children. Sandel's solution is as reasonable as it is vague; he proposes that parents strike a more even balance between accepting love and transforming love in their approach to their children's environment.

But if accepting love and transforming love can be balanced in the environmental domain, why can they not also be balanced in the genetic domain? How can it be that moderate transforming love is an appropriate norm in the case of the child's environment but an inappropriate norm in the case of her biological nature? In sum, we may pose to Sandel the same question we posed to O'Donovan: Why does he not recommend the same moderation in the approach of parents to their children's biological characteristics as he recommends in their approach to their children's environment? Like O'Donovan, Sandel fails to consider the possibility that genetic and environmental interventions are subject to the same moral principles. In the end, he simply reasserts his point that "eugenic parenting" is problematic because it "fails to appreciate the gifted character of human powers and achievements." But the question is whether an

---

[69] Ibid., p. 52. He borrows the term from a book of that title by Alvin Rosenfeld and Nicole Wise.

appropriate appreciation of these powers and achievements, as an expression of accepting love, is compatible with moderate efforts to enhance them, as an expression of transforming love, and Sandel's summary rejection of genetic enhancement begs that question.

God, Nature, and the Restraint of Human Mastery

Like Habermas, Sandel ventures a brief experiment in theology, and it is worth our while to supplement this examination of his version of NS1 by considering how that experiment illuminates the principal shortcoming of his position on the biotechnological enhancement of children by their parents. Commenting on a theme in the work of the Jewish theologian David Hartman, who is correcting what he sees as a problem in the work of his renowned teacher, Rabbi Joseph B. Soloveitchik, Sandel describes what he sees as an ambivalence regarding human mastery of nature in Judaism.[70] On the one hand, the sharp distinction between the Creator and the creature to which Judaism is committed authorizes a permissive stance toward nature, justifying the subjection of the latter to human control and use. There is a strong denial that God is one with nature or is embodied in nature, and this denial is generally taken to imply that there should be little if any restraint in principle on human control and mastery of nature – a position that was taken by Soloveitchik and is echoed by many contemporary Jewish bioethicists. On the other hand, the sharp distinction between the Creator and the creature is also the occasion for humility before God. But, Sandel explains, whereas Soloveitchik found this humility in submission to the inscrutable will of God as exemplified by Abraham in the *akedah* (that is, the binding of Isaac narrated in Genesis 22), he thinks – in contrast to most Jewish bioethicists – that Judaism provides grounds for finding humility in the observing of limits on the exercise of human power over nature.

Sandel wants to show how permissibility and restraint with respect to nature are compatible with each other, comprising two aspects of a single moral and theological vision that restrains human mastery within a general affirmation of human autonomy, initiative, and creativity in relation to nature. To combine the two stances, he appeals to the distinction between therapy and enhancement, proposing that there be no restrictions in principle on human mastery of nature for the purpose of restoring human

---

[70] In what follows I will consider Sandel's argument on its own terms without considering whether his interpretations of Soloveitchik and Hartman are adequate.

beings to wholeness in the face of disease but that the power to design and select biological characteristics be reserved for God alone.[71] I have noted several times in this chapter the disagreement over the viability of the distinction between therapy and enhancement, which may not support the sharp line Sandel wants to draw between permissibility and restraint. But what is more interesting in the present context is his theological argument for restraint in the case of enhancement. For him, what is theologically at stake in biotechnological enhancement is not anything to do with nature or even with the human being whose nature it is, but rather an illicit human self-deification. "The limits on the exercise of human powers over nature arise not from nature itself but from a proper understanding of the relation between human beings and God. If it is wrong to clone ourselves in a quest for immortality, or to genetically alter our children so that they will better fulfill our ambitions and desires, the sin is not the desecration of nature but the deification of ourselves."[72] What is problematic about the intentional alteration of nature is that by it, humans transgress the boundary between creature and Creator in "a hubristic quest to usurp God's role."[73] To take nature into our hands is to exercise mastery and dominion that belong to God alone.

Sandel's worry about usurpation of a divine prerogative situates him in the space of the Prometheus myth and thus invites the objection that he has left the space of Jewish theology. However, what is notable for my purpose is that nature in this account serves as a mere proxy for a negative obligation (namely, nonusurpation) of the creature to the Creator. Nature is inviolable not because it is God's property (in which case it would be sacred in the strict sense), nor (as with O'Donovan) because as God's creation it is good and thus deserves respect, nor even (as with O'Donovan and Habermas) because it is wrong for humans to assert themselves as God toward their fellow human creatures by intervening into their nature. While O'Donovan and Habermas (in the tentative theological venture mentioned previously) argue that humans should not presume to act as creators toward their fellow humans, Sandel argues that humans should not assume for themselves God's power. Human biological nature is inviolable because there is a religious requirement to restrain mastery. The fact that it is human nature that is declared off-limits is entirely secondary, if not arbitrary. What matters is that a limit to human mastery

---

[71] Sandel, "Mastery and Hubris in Judaism," pp. 202f., 210.
[72] Ibid., p. 201.
[73] Ibid.

is required; that it is human nature that constitutes this limit is almost beside the point.

It is striking that this same logic governs Sandel's treatment of biotechnological enhancement in his nontheological work. Parenthood requires restraint of the urge to control; parents must therefore leave the biological nature of their children untouched. This view of the normative status of human nature, or at least this version of it, fails because Sandel gives no reasons to believe that either piety or parenthood require total restraint of mastery, or to believe that human nature, and specifically the genotype of one's child, is the appropriate object of that restraint. Yet the requirement of total restraint remains in force, and it is, according to Sandel's version of NS1, the ultimate ground for attaching normative status to human biological nature as that which simply confronts us. Once that requirement is called into question or qualified, the case for leaving human nature off-limits to our intervention vanishes or is qualified as well.

Conclusion

According to NS1, normative status attaches to human nature considered as that which is, and is what it is, apart from our willful activity. As such, NS1 declares human nature to be, at least in principle, off-limits to attempts to determine its biological characteristics. This is the fundamental claim of NS1. As we have seen, this claim does not imply that human nature is untouchable because it is sacred, that we have access to human nature without the mediation of culture, or that human nature is not in part the product of human activity. It need imply only that the biological characteristics of others should be kept off-limits to our attempts to determine them because those attempts involve one or another form of illicit power over those others (though it may also imply that such interventions involve a problematic stance toward human biological nature).

For NS1, the person is the proper bearer of normative status. Whether the normative status of the person is protected by the created order in which human persons are generic equals (O'Donovan), by her status as a moral and political subject (Habermas), or by institutions and practices such as parenthood (Sandel), our three authors agree that the normative status that attaches to persons extends to their biological nature (or at least, in the cases of Habermas and Sandel, to the biological nature of children) in such a way that the latter is, at least in principle, to be granted immunity from determination by intentional human action. In this sense, they all adhere to NS1, according to which normative status attaches (in their cases, indirectly or by extension) to human nature as that which exists, and is what it is, apart from intentional determination. Yet we have seen that all three versions of NS1 fail to secure the claim that human biological nature should be granted immunity from attempts to determine it. In O'Donovan's case, acting and making are not easily separable, while parenthood stands in multiple generic and teleological orders, some of which seem to rule out biotechnological enhancement of children by their parents, and some of which seem to permit it. In Habermas's case, it is reasonable to suppose that at least some choices parents make for their children would be endorsed by the latter, and the asymmetries these choices involve are for the most part not any more pernicious in principle than those that are involved in childrearing. In Sandel's case, there is no reason to suppose that accepting love and transforming love, both of which are essential to good parenting, cannot be balanced in parental interventions into a child's biological nature as they presumably can be in parental interventions into a child's environment. In sum, in none of these cases is it clear that biotechnological determination of human nature in principle involves illicit power over those whose nature is selected, altered, or controlled. We must conclude that the strongest versions of the most plausible arguments made by the most competent proponents of NS1 are unable to vindicate the defining claim of NS1 regarding the moral inviolability of human nature in the face of its determination by biotechnology. If it is wrong to determine human biological characteristics, it is not because normative status attaches to human nature as that which is, and is what it is, apart from our willful activity.

However, it does not follow from this conclusion that NS1 is entirely mistaken or that it is irrelevant to the ethics of biotechnological enhancement. Although NS1 fails to justify its claim that human nature should be off-limits to biotechnology, it accounts for a danger that always accompanies the determination of the biological characteristics of children. The selection or design of the biological characteristics of children inevitably involves a morally problematic comportment toward them, whether with O'Donovan we characterize it as making, with Habermas as engaging with the child as an object rather than a fellow subject, or with Sandel as treating the child as a project that may succeed or fail. That these problematic comportments toward the child also occur in the determination of the child's environment does not mitigate their seriousness. In the end, as we have seen, NS1 fails to rule out the possibility that at least in some instances or to some extent these comportments may simply be potentially dangerous features of acts that meet a genuine duty of parents to benefit their child and may thus be capable of being justified in principle. That is clearly the case with many environmental interventions, and we have seen that NS1 fails to exclude the possibility that it may also be the case in at least some biological interventions.

At the same time, NS1 exposes dangers that inevitably accompany the selection or design of biological characteristics, and it is unclear how these dangers can be recognized and avoided without the awareness, which NS1 promotes, that children are to be treated as equals who are begotten by us, not inferiors who are made by us; as fellow subjects even as we act on them; and as recipients of unconditional love even as we practice transforming love along with accepting love. Recognition that their biological nature is not at our disposal is obviously crucial to this awareness, and the nondisposability of their nature can, in my view, be established on any one of these grounds or on all three of them. This recognition would, also in my view, be in principle compatible with some acts of determining the biological characteristics of others if (and only if) there are convincing reasons for holding that those acts discharge a broad duty of parents to benefit their children. If and when an intervention counts as discharging a duty to a child, what is due to the child constrains the action of the parents, and we should therefore not count it as an instance of putting the child's nature at their disposal. Whether any acts of determining the biological characteristics of children may count as discharging a duty to benefit a child and what acts these might be I leave to the following chapters and ultimately to the ethical evaluation of particular biotechnological enhancements. For present purposes, however, it is important to close by asserting that because it formulates the principle that the biological nature of others is not at our disposal, NS1 remains indispensable to the ethics of biotechnological enhancement and should be included in any approach to the latter that is undertaken in Christian ethics.

CHAPTER 3

Human Nature as Ground of Human Goods and Rights

In many debates over biotechnological enhancement, normative status is claimed for human nature as the ground of distinctively human goods or rights. There are many ways to assert this claim, but the most characteristic formulations of it are broadly Aristotelian: They hold that to speak of the nature of a living thing is to refer to the characteristics or processes that make it what it is (that is, its essence), while its good consists in activities, capabilities, states, or conditions in which it thrives as the kind of thing it is and (at least for rights-bearing beings) rights protect its good, so understood. I will refer to this position as NS2. From the standpoint of Christian ethics, NS2 is attractive for two reasons. First, its connection of goods and rights with human nature as created by God both supports the Scriptural claim that creation is good (our creaturely nature is ordered to our flourishing and is not indifferent or hostile to it) and provides an intelligible ground for goods and rights (goods are not merely preferences we happen to have and rights are not merely useful inventions but both are related to our nature as God created it). Second, the connection of human goods with human nature provides a basis for judgments about our good. By looking to our nature, we can determine what is and is not genuinely good for us.

The prospect of biotechnological alteration of human biological functions and traits poses a challenge to all versions of NS2. As long as our ability to affect these functions and traits is severely limited, it can be taken for granted that genuine human goods are those that fulfill human nature in its present state and that human rights protect the pursuit of those goods. But what will happen if we become capable of making significant changes to our biological nature? In that case, human nature becomes a variable rather than a constant. To be sure, any such capability is likely to be modest in the foreseeable future, and the extent to which human nature is a variable rather than a constant will be accordingly modest. Nevertheless, NS2 has until now assumed that human nature in its present form is the ground of human goods and rights, and the alterability of human nature by biotechnology poses questions of what it means for NS2 that this ground might shift. This chapter considers two such questions. First, if human goods and rights depend on human nature, and if biotechnology alters human nature, are those goods and rights imperiled? In other words, does the alteration of human nature by biotechnology threaten human goods and rights, which (it is held) require human nature to hold stable or to remain in its current state? Second, if biotechnology eventually becomes capable of altering our nature in accordance with what we consider to be good, what follows for NS2 as a claim about the normative status of human nature? NS2 holds that in some way we look to our nature to determine what our good is; we do not look to whatever we might consider to be good to determine what our nature will be. But if we do become capable of determining what our nature will be based on what we consider to be good, won't we be torn between (1) attaching normative status to our nature, at the expense of goods that require alteration of our nature, and (2) attaching it to what we consider good, at the expense of our nature? And won't this situation discredit the claims that our nature is ordered to our good and that our good is intelligible in terms of our nature?

These two questions are pressing in light of two typical moves made by the most capable subscribers to NS2. With these moves, the plausibility of NS2 has been successfully defended against objections, made on scientific grounds, that it is incompatible with evolutionary and environmental accounts of human nature, and on philosophical grounds, that its derivation of goods and rights from human nature is conceptually untenable and ethically problematic. First, these subscribers to NS2 argue that the observable variation and change that characterize human nature are compatible with the most defensible versions of Aristotelian essentialism, for which membership in a kind (such as the human kind) does not require rigid uniformity or permanence in the functions, traits, and behaviors that constitute the kind. Members of a kind may exhibit variety, and their behaviors and traits may undergo change over time, all without threatening their identity as members of that kind. This accounts for how human beings of widely divergent cognitive, emotional, and physical characteristics still count as human. Second, they argue that human goods are not inherent in human nature, such that we can directly identify them from descriptions of human nature, but rather fulfill human nature in ways that are not strictly determined by human functions and traits. The first move seems to imply that kinds of things (in this case, the human kind) are flexible enough to accommodate whatever variation and change biotechnology may introduce (just as they accommodate the variation and change that is due to evolutionary and environmental causes). The second move distances the human good from human nature in a way that seems to permit the alteration of human nature in accordance with what is considered to be good: If our good is not strictly derivable from our nature anyway, why not alter our nature in accordance with whatever we think is good? In both cases, the distinctive claim of NS2, namely, that human nature is the ground of human goods and rights, so that we look to the former to determine what the latter are, seems to be compromised. The very moves that enable NS2 to defeat scientific and philosophical objections to its key claims seem to render it vulnerable in view of the prospects of biotechnological alteration of human nature. It is not clear, in view of these prospects, how human nature can be affirmed as the ground of human goods and rights and how normative status can attach to human nature with respect to that role.

This chapter considers the challenge that is posed to NS2 by these questions regarding the threat of biotechnology to human goods and rights and the claim that we look to our nature to determine our good and not vice versa. Two opponents of biotechnological enhancement, namely, Francis Fukuyama and Leon Kass, have addressed these questions at length, albeit not in the precise form in which I have just posed them. Their versions of NS2 are discussed subsequently. However, NS2 has stakeholders who do not directly participate in debates over biotechnological enhancement. Among them are Martha Nussbaum and Jean Porter, whose views of the normative significance of human nature can be seen as versions of NS2, though not necessarily as positions on biotechnology. In any case, their views are relevant to our two questions and will also be considered in what follows. But before we turn to these questions, it is necessary to clarify what NS2 means by its claim that human goods and rights are grounded in human nature.

Goods, Rights, and Human Nature

As with NS1, false or superficial characterizations of NS2 are prevalent in the bioethics literature, so once again I begin by rejecting these characterizations and formulating a plausible version of NS2. Critics of NS2 tend to treat the claim that goods or rights are grounded in human nature as a trivial claim or to identify it with the strong claim that goods or rights are inherent in features of human nature. However, the claims made by sophisticated versions of NS2 are neither as weak nor as strong as these criticisms suppose. So, before turning to the questions biotechnological enhancement poses to NS2, I will clarify the claims about the relation of human goods or rights to human nature made by more and less sophisticated versions of NS2.

Human Nature and the Human Good

First, what is meant by the claim that human *goods* are grounded in human nature? A weak or trivial version of this claim merely reminds us that human nature places constraints on what may reasonably count as human goods. Consider, for example, a sophisticated desire- or preference-satisfaction account of the good according to which one's good is not simply what one consciously desires or wants but what one would desire or want if one were adequately informed about all the factors relevant to one's desires or wants. Knowledge of the parameters that human nature sets for the fulfillment of one's desires or wants surely counts as one condition of being adequately informed about them. It would not be reasonable to orient one's desires or base one's projects on, say, the ability to work without sleeping, which one might find attractive but for which one's natural capacities, or human capacities generally, are unsuited. This version of the claim that human goods are grounded in human nature is a weak one. It does not hold that goods fulfill our nature but only that they are constrained by our nature.

In contrast to this weak version is a strong version that claims that human goods are inherent in human nature and that moral norms can therefore be deduced from descriptions of human nature. Some older and now discredited versions of NS2 treated human nature as an internal teleological system in which particular biological functions, such as sexual functions, are ordered to determinate ends, such as procreation, which fulfill those functions, while moral norms, such as sexual norms, are derived directly from the relation of a function to its end. This view treats human biological nature somewhat as a mechanical artifact, and it follows from it that human goods are inherent in human biological nature (that is, they are realized by actions that respect the ordering of functions to their determinate ends), while the determinacy of the ends (that is, their specificity and the fact that they can be directly read off the functions to which they are related) makes it possible to derive substantive moral norms directly from human nature.[1]

---

[1] This position, which its critics call "physicalism," was prominent in pre–Vatican II Catholic moral theology.

As I noted, critics of NS2 tend to assume that these weak and strong versions are the only alternatives. Not surprisingly, they therefore find it easy to dismiss NS2.[2] The weak version, after all, is too trivial to count as a genuine instance of NS2 (to say that the good is constrained by human nature is not to say that it fulfills human nature), and the strong version is problematic, both because it tries to deduce normative statements from descriptive statements and because it breaks our nature up into discrete functions rather than treating it as a whole. I concur with these assessments and will not consider the weak or strong versions of NS2 in this chapter. However, between these weak and strong versions lies a moderate (and more plausible) version that is typically overlooked or unrecognized in bioethical debates. This version is found in a variety of forms, two of which are important for this chapter. One form, which is represented in this chapter by Jean Porter, holds that there are natural human goods that, while not inherent in human biological nature, fulfill human biological nature (understood as a complex whole, and not just in terms of particular functions and traits) in specific ways, while extranatural factors play a major role in determining concretely what does and does not count as fulfilling it. To take a straightforward example, human beings are by nature social, so that some form of social existence is necessary for their fulfillment. But because our social nature is indeterminate, many particular social arrangements can provide at least adequate fulfillment. It follows that the norms by which we judge which of these alternatives best instantiates the good cannot be derived from a description of our nature, because our nature, taken alone, would support any of them. We must therefore arrive at these norms through practical reason reflecting on rational principles, historical experience, and local circumstances, and not directly on our nature.[3]

Rather than limiting the role of our biological nature in the determination of norms, the second form of this moderate version of the claim that human goods are grounded in human nature expands what we understand our biological nature to be. On this view, human biological nature encompasses the ways in which biological characteristics that are shared with other living things (such as mortality and sexual reproduction) take distinctively human form in desires, aspirations, and social bonds in which

---

[2] See, e.g., Allen Buchanan, *Beyond Humanity? The Ethics of Biomedical Enhancement* (New York: Oxford University Press, 2011).

[3] See Jean Porter, *Nature as Reason: A Thomistic Theory of the Natural Law* (Grand Rapids, MI: Eerdmans, 2005).

the characteristic form of life of the human organism is lived.[4] The most prominent contemporary representative of this attempt is Leon Kass, who calls for "an ethical account of human flourishing based on a biological account of human life as lived, not just physically, but psychically, socially, and spiritually."[5] Rather than stressing, as Porter does, how indeterminate requisites of our biological nature are made determinate in historically and socially particular forms by the exercise of practical reason, Kass stresses how reflection on basic aspects of our biological nature discloses meanings, longings, and forms of attachment that constitute a distinctively human biological life. For example, upon reflection, the distinctively human longing for perpetuation through offspring and readiness to devote a life to a cause are seen to be characteristic of a form of life that must consciously come to terms with the mortality it shares with other living things. Longings and attachments like these constitute the distinctively human form of life as one that is characterized by conscious struggle with and partial transcendence of biological necessities that are endemic to living things generally.

In sum, Porter limits the role of our biological nature in determining our good, while Kass expands our notion of our biological nature to encompass the form of life that is constituted by engagement with biological nature in the narrower sense. But in spite their differences, both Porter and Kass subscribe to a moderate version of the claim that human goods are grounded in human nature. That version, which is the exclusive focus of this chapter, goes beyond the trivial sense in which the weak version relates human goods to human nature while avoiding the problems associated with the strong version.[6]

---

[4] See Hans Jonas, *The Phenomenon of Life: Toward a Philosophical Biology* (New York: Harper and Row, 1966); and Leon Kass, *Toward a More Natural Science: Biology and Human Affairs* (New York: The Free Press, 1985).

[5] Leon Kass, *Life, Liberty and the Defense of Dignity: The Challenge for Bioethics* (San Francisco: Encounter Books, 2002), p. 21.

[6] A third version, which I do not discuss here because it is presented as an intervention in a long debate in metaethics that is outside the scope of this book, is that of Philippa Foot, whose argument in *Natural Goodness* (Oxford: Oxford University Press, 2001) grounds the goodness of human actions and dispositions in aspects of human nature as a living form (or species). According to Foot, "[T]he fact that a human action or disposition is good of its kind [is] simply a fact about a given feature of a certain kind of living thing" (p. 5). E.g., the goodness of promise keeping as a kind of action is a fact about humans as a kind of living thing for which the cooperation of other members of its kind is a necessity, just as, e.g., having water is a necessity for plants and building nests is a necessity for birds (pp. 15f.). Foot goes on to ground norms in these natural necessities, defining goodness and defect in terms of kinds of behavior that instantiate or fail to instantiate these necessities. Thus far, Foot's version resembles the strong version dismissed previously: Moral norms are directly derivable from human nature. However, she stresses that in the human kind of living thing, natural necessity is highly complex and diverse, rendering it impossible to deduce determinate norms from natural

Human Nature and Human Rights

Let us now briefly (and rather inadequately) consider the claim that human *rights* depend on human nature. Once again, there are multiple versions of this claim, but two are commonly invoked in debates over biotechnological enhancement. According to a strong version, rights are thought to in some way inhere in human nature by virtue of certain human characteristics, such as consciousness or emotion, or to supervene on those characteristics. To the extent that these characteristics have a biological component or depend on human biological nature, certain qualitative or quantitative alterations to human biological functions or traits could alter the content of the relevant rights or their conditions of applicability.[7] The point in either case is that on this account, a being with a significantly different form of consciousness or different emotions might well be the subject of different rights, or perhaps of none at all. By contrast, a moderate version of the claim holds that rights are means by which societies recognize and protect needs, interests, or goods that are deeply rooted in human nature and are thus necessary to human flourishing. This moderate version is distinguishable from weak versions for which rights are mere constructs that are politically useful or necessary. It holds that although rights are conferred by society, and are in that sense not natural, they nevertheless answer to certain requirements of human nature and should therefore be considered as natural rights. In this chapter, this moderate position is represented by Francis Fukuyama.[8]

Biotechnology and Natural Goods and Rights

The defining claim of NS2 is that human goods or rights are grounded in human nature, so that we look to our nature to determine what our goods and rights are. It should now be clear that sophisticated versions

---

necessities as one can do in the case of other living things (and as the strong version previously does in the case of humans) (pp. 42f.) while nevertheless preserving the fundamental claim about the grounding of norms in human nature. In relevant respects, then, her position is similar to Porter's.

[7] Of course, the likelihood of total inapplicability seems to be low if, as is often the case, characteristics such as consciousness and emotion are said to ground rights. It is unlikely that biotechnological enhancement will result in the kind or degree of change to these characteristics that would imperil any rights that are grounded in them. It is more likely that radically different cognitive or emotional capacities than humans now have would involve the ascription of different rights or, more plausibly, different specifications of the same rights.

[8] See Francis Fukuyama, *Our Posthuman Future: Consequences of the Biotechnological Revolution* (New York: Farrer, Straus, and Giroux, 2002) pp. 101–02, 110–11, 125, 127–28.

of NS2 do not hold that these goods are deducible from a description of human nature (the strong version), on the one hand, or that they merely take account of human nature but do not in any way fulfill or protect it (the weak or trivial version), on the other hand. It should also be clear that these versions of NS2 may hold either that rights are inherent in human nature or supervenient on it (the strong version), or that they are conferred in view of certain requirements of human nature (the moderate version), but not that they are merely politically necessary or expedient constructs (the weak version).

We are now in a position to consider the two questions, noted previously, that biotechnology poses to these sophisticated versions of NS2. First, would biotechnological enhancement imperil certain goods or rights that, according to NS2, depend on human nature in its current, unenhanced state? Second, if biotechnology develops to the point that we can begin to determine what human nature will be in accordance with what we consider to be good rather than (as NS2 has it) looking to human nature to determine what is good, then does NS2 dissolve? In the face of these questions, subscribers to NS2 appear to have sound reasons for opposing at least some biotechnological enhancements (as most subscribers to NS2 who consider biotechnology do). Yet the bar for opposition to biotechnological enhancement in the case of NS2 is considerably higher than it is in the case of NS1. For NS1, we recall, normative status attaches to human nature as that which is, and is what it is, apart from our intentional determination of it. *Any* intentional effort to select or alter biological characteristics will violate the normative status of human nature in that sense. For NS2, however, not just any effort to determine biological characteristics in these ways will violate the normative status of human nature, but only those interventions (whatever they might be) that endanger the human goods or rights that depend on those characteristics or that aim at goods that are both genuinely good and cannot be reasonably explained as fulfilling our nature as it now is. To oppose biotechnological enhancement on the first ground, one must show how the relevant interventions imperil natural goods or rights. To oppose it on the second ground, one must show either how the goods that biotechnology makes available to us are not genuine goods or that they can only be explained as the fulfillment of something other than in our nature as it now is. We will see that these are not easy conditions to meet. They can be met only by rebutting two arguments that are made by proponents of biotechnological enhancement.

First, proponents of biotechnological enhancement deny that biotechnology endangers nature-grounded goods or rights by insisting that, with respect to these goods or rights, there is no relevant difference between what biotechnological enhancement does to human nature and what nature has done to it through evolutionary and environmental causes.[9] According to their argument, if human goods and rights are not imperiled by the variation and change that nature visits on human biological functions and traits, then there is no reason to worry that they will be imperiled by the variation and change that biotechnological enhancement might bring. They point out that apart from biotechnology there is already considerable variation and change in human functions and traits, yet this already-existing variation and change do not seem to imperil human goods and rights. Why, then, should we suppose that variation and change introduced by biotechnology will imperil them?

Second, proponents of biotechnological enhancement argue that normative status properly attaches to goods or rights as such, and not to the human nature on which they allegedly depend, so that it is justifiable in principle to alter human nature to protect or promote these goods or rights. Up to now, of course, the goods and rights we enjoy and affirm are those that fulfill and protect our nature as it is. But unless we think that goods and rights are inherent in our nature (a view we rejected as unattractive in the case of goods), we can, as it were, abstract from our nature and ask whether it is adequate to our good and our rights as we understand them. If it then becomes possible to remake our nature, we can do so to make it more adequate to these goods and rights. And if it is justifiable to do that, it is because normative status properly attaches to those goods or rights, and not to the nature on which they depend.

In short, in response to the first question (If human goods and rights depend on human nature, are they imperiled by the biotechnological alteration of human nature?) proponents of biotechnological enhancement argue that there is no relevant difference between natural and technological changes to human nature, while in response to the second question (Is NS2 lost in the choice between looking to our nature to determine our good and looking to what we consider our good to determine what our nature will be?) they argue that normative status attaches to our good and not to our nature. The rest of this chapter considers these two arguments.

[9] Buchanan, *Beyond Humanity?*, p. 120. I refer to environmental as well as evolutionary causes of variation and change to account for phenomena such as increased stature and longevity that are not due to natural selection.

Does Enhancement Differ from What Nature Does?

We begin with the first argument. NS2 presupposes that human biological nature prior to biotechnological alteration is stable enough to ground determinate and enduring human goods or rights. But we know that due to evolutionary and environmental factors – and thus prior to biotechnological intervention – human characteristics are in flux and vary significantly from one individual to another. It might seem reasonable to infer from this commonly recognized fact that natural human goods or rights must likewise be changing and variable. If human nature as the sum of these characteristics and their interrelations is instantiated differently from one individual to another, both synchronically and diachronically, must not the goods and rights that are grounded in it likewise differ from one to another? Subscribers to NS2 reject this inference and deny that goods or rights are subject to this kind of diversification. As they see it, these goods or rights are not so rigidly dependent on human nature that they cannot accommodate the variability and change caused by evolutionary and environmental factors. The ground of this compatibility of natural goods or rights with the variability and change of human nature may be that human nature underdetermines the relevant goods, which are concretely realized in particular social forms and in accordance with individual abilities, proclivities, and circumstances. In this case, the specific realization of goods in the lives of individuals accords with the variability among individuals. This kind of compatibility is implied by Porter's position.[10] Alternatively, the compatibility may obtain because the natural characteristics on which the relevant rights depend need not be rigidly fixed or invariable to support them. As long as the relevant characteristics fall within a certain biostatistical range, human nature is sufficiently uniform to support the ascription of the same rights to every human being. This kind of compatibility follows from Fukuyama's position.[11] Finally, as Kass's position would have it, the compatibility may obtain because the relevant goods require only the presence of certain very basic features of biological nature, such as sexual reproduction and mortality. Considered simply as such, these features have remained stable and invariant across many evolutionary ages and biological species, and the relevant human goods require no more than the mere incidence of these features in human nature.[12] But whatever

[10] Porter, *Nature as Reason*, pp. 53–139.
[11] See Fukuyama, *Our Posthuman Future*.
[12] See Kass, *Life, Liberty and the Defense of Dignity*.

the explanation these versions of NS2 offer for the compatibility of human goods or rights with the change and variability of human nature, it follows from all of them that human nature need not be rigidly fixed or invariant to ground the stable, enduring goods or rights that many subscribers to NS2 seek to protect.

In sum, nature-grounded goods and rights can accommodate a range of change and variation in the human nature in which they are grounded. However, if we acknowledge that the nature in which human goods and rights are grounded is changing and variable, two questions arise. First, is there any stable condition of human nature for biotechnology to interrupt? And second, if subscribers to NS2 concede that the variation and change introduced by evolutionary and environmental factors pose no threat to the stability of human nature and the goods and rights that depend on it, then why should they worry that this stability is threatened by the variation and change that biotechnology could introduce? The two following subsections address these two questions in order.

Biotechnology and the Stability of Human Nature

First, opponents of biotechnological enhancement who appeal to NS2 typically presuppose that biotechnological enhancement introduces changes to a nature that is otherwise stable. But is their presupposition of a stable human nature tenable in light of biological evolution? The most widely accepted theories of a stable human nature are Aristotelian.[13] Classical versions of these theories hold that human nature is a natural kind defined by a characteristic or a set of characteristics that all humans and (at least in their distinctively human versions) only humans share. The Darwinian discovery that species are evolved products of contingent evolutionary processes is thought by critics of Aristotelian theories to have invalidated the notion of unchanging essences shared by all humans and only humans. However, Porter points out that our *concepts* of species amount to more than mere generalizations about their diverse individual members and do not share the contingency of the processes that have

---

[13] There is a large and impressive literature on Aristotelian and neo-Aristotelian biology. Because my topic is not theories of nature or human nature as such but claims that normative status attaches to human nature, I have limited my discussion in this subsection to two neo-Aristotelians who make that claim (albeit not in my explicit terms) and elaborate it at length, namely, Jean Porter and Francis Fukuyama.

produced species. First, even in a post-Darwinian context, she argues, we can explain anomalous or changing traits or behaviors of an organism only if we can determine the kind of organism it is and can understand the unusual or new trait or behavior as a possibility of creatures of that kind. Second, we can explain the development of organisms and debate whether a disputed member is one of them or not only by appealing to the notion of a fully formed or paradigmatic member of their kind, to which less developed organisms or organisms that exhibit anomalous traits approximate in varying degrees.[14] Of course, not all philosophers of biology would concede that biological explanation requires (or even allows for) these appeals to forms and ends. However, these two cases demonstrate, respectively, how the basic Aristotelian understanding of essences as formal and final causes can accommodate the variation and change emphasized by post-Darwinian biological science without surrendering the stability of the natures of things.

Fukuyama seeks to vindicate a fundamentally Aristotelian position by way of a biostatistical rather than a metaphysical conception of human nature, defining the latter as "the sum of the behavior and characteristics that are typical of the human species, arising from genetic rather than environmental factors."[15] By "typical" Fukuyama has in mind a statistical norm that allows for variability and change (and is thus compatible with evolutionary biology). Traits and behaviors count as components of human nature (as distinct from nonessential characteristics) if they are genetic in origin and their distribution can be represented on a bell curve with a median point and a relatively small standard deviation. This model accommodates variation in its acknowledgment that the distribution of many traits (for example, eye or skin color) will fall outside the standard deviation and thus will not count as constitutive of human nature, and it accommodates change in its acknowledgment that the median point for traits may shift (as, for example, median height and life expectancy did in twentieth-century Europe and North America). Once again, neither the biostatistical approach nor Fukuyama's version of it are universally credited by philosophers of biology, but it does demonstrate how a broadly Aristotelian position accommodates variation and change without surrendering stability.

[14] Porter, *Nature as Reason*, pp. 82–125.
[15] Fukuyama, *Our Posthuman Future*, p. 130.

Is Biotechnology a Threat to Human Goods and Rights?

Both Porter and Fukuyama, then, provide subscribers to NS2 with the means to defeat the objection that there is no stable human nature for bio-technological enhancement to unsettle. Porter's metaphysical forms and Fukuyama's statistical norms avoid the problems that biological evolution poses to more rigid versions of Aristotelian essentialism and thus seem capable of stabilizing the concept of human nature in a way that grounds natural goods and rights while also accommodating variation and change.[16] However, we now face the second question posed by proponents of biotechnology: Can a concept of human nature that is capable of accounting for the observable variation and change among human functions and traits provide a basis for ruling out the variations and changes to these functions and traits that biotechnology promises to introduce? If human nature is already variable and changing due to natural causes, and that without any apparent threat to nature-grounded goods and rights, then why should we worry that the variation and change introduced by biotechnology will imperil those goods and rights?

A subscriber to NS2 may respond by denying that biotechnologically introduced variation and change pose any problem at all here. Porter does not discuss the prospect of biotechnological alteration of human nature or its implications for her position, but there is no indication that she would find biotechnologically introduced variation and change to be any more troubling than what has been introduced by evolution. If that is so, then NS2 can readily accommodate biotechnological enhancement in principle. Fukuyama, however, discusses this issue at length and argues that variation and change caused by biotechnological enhancement does

---

[16] Porter and Fukuyama rely on the claim that metaphysical forms (Porter) or biostatistical norms (Fukuyama) are paradigms, i.e., ideal or fully realized states of functions or traits on the basis of which we can identify variant instances of the same function or trait and determine their degrees of approximation to the ideal or fully realized state. On this basis forms or norms supply a ground for identifying genuine goods or rights while also accommodating variations and changes in human functions and traits. The assumption that forms or norms can be thought of as paradigms is questioned by the line of thought initiated by Georges Canguilhem, who argued that physiological norms are not natural givens but merely the provisional effects of the norm-generating capacity of biological life in specific natural (and for human beings, also social) environments. I say more about Canguilhem's account in Chapter 4, but if it is correct, then there are no paradigmatic forms or norms that concepts or bell curves represent, and proponents of biotechnological alteration of human nature will be able to continue to argue that, at least biologically speaking, there is no fundamentally stable human nature for biotechnology to disrupt. For the sake of the argument of this chapter, however, I assume that Porter's forms and Fukuyama's norms can be vindicated in the face of Canguilhem's arguments. See Georges Canguilhem, *The Normal and the Pathological*, translated by Carolyn R. Fawcett in collaboration with Robert S. Cohen (New York: Zone Books, 1991).

pose a unique threat. His version of NS2 is therefore vulnerable in the face of the question regarding the difference between the variation and change introduced by evolution and that introduced by biotechnology.

Fukuyama approaches biotechnological enhancement as a political theorist. He argues that liberal democratic orders have proven to be more stable than their recent competitors because they are more consonant with human nature, by which he means that they are based on rights that pick out and protect needs and interests that are deeply rooted in our nature.[17] But if rights are related to human nature in this way, their ascription to human beings as such (that is, their universality) presupposes that all human beings share the needs and interests that these rights pick out and protect, which is to say that they share a common nature. It also presupposes that human beings mutually recognize one another as entitled to equal respect by virtue of their common nature. It is here that Fukuyama thinks biotechnology threatens liberal democratic orders. He worries that some biotechnological enhancements, such as those that could result in wide gaps in cognitive capacities or ranges of mood (consider, for example, proposals to eliminate shyness and aggression) between those who do and do not have access to the relevant technologies, may destroy the common nature that is the basis of rights and threaten the natural equality on which mutual recognition is based.[18]

The points we have just considered about variation and change suggest that this worry is exaggerated. If the biological nature of unenhanced humans is currently stable enough to serve as the ground of human goods or rights despite its variability among individuals and its changes over time, why should we suppose that biotechnologically enhanced human nature will imperil these goods and rights with the variation and change it introduces? Consider once again the variation and gradations among groups brought about by the increases in stature and longevity in twentieth-century Europe and North America or by the effects of literacy on brain functioning. At least in principle, we not only recognize those

---

[17] See Fukuyama, *Our Posthuman Future*, pp. 101–2, 110–11, 125, 127–28.

[18] Ibid., pp. 153–60. This worry appears to be directed at three possible scenarios that Fukuyama never clearly distinguishes or identifies as such: (1) that alterations of human nature will diversify needs and interests to the extent that the rights that pick out and protect these needs and interests will no longer be universal; (2) that divergences between the capacities of enhanced and unenhanced individuals may become great enough to amount to gradations within the human species, thus undermining natural equality, mutual recognition, and the rights that rest on equality and recognition; and (3) that the very fact that human nature has been rendered variable and manipulable will undermine the stability of a common human nature to which universally recognized rights attach. See Fukuyama, *Our Posthuman Future*, pp. 125, 127–28, 154–57, 171–74.

who have not undergone these dramatic changes as sharers with us in a common biological nature, we also often acknowledge a broad duty to reduce the resulting gradations by making the changes we have undergone available to them.[19] It seems reasonable, then, to assume that we would, at least in principle, recognize people with significantly enhanced cognitive abilities or different intensities or ranges of emotion as sharers in our nature, to say nothing of people with even greater height or life expectancy than ours.[20] In short, even if we agree that there is a stable biological human nature, any metaphysical form or biostatistical norm that is flexible enough to accommodate the variation and change that characterize the human species in its unenhanced state without imperiling either our concept of human nature or the goods or rights that depend on it will also permit a wide range of biotechnological alterations without incurring these threats.

To be sure, this range is not unlimited. At some point, on either Porter's account or that of Fukuyama, quantitative differences in the same characteristic or qualitative differences such as a different range of social emotions or entirely new cognitive or perceptual functions (such as perceiving other dimensions) could be significant enough to imperil natural equality or the recognition (and indeed the existence) of a common human nature and imperil whatever rights depend on the latter. But we may safely assume that the range accommodated by Porter's forms or Fukuyama's norms is wide enough to encompass the biotechnological enhancements that currently appear to be realistically achievable. To that extent, the proponents of biotechnological enhancement seem to have defeated the claim that biotechnological enhancement threatens nature-grounded goods or rights by arguing that, with respect to these goods or rights, there is no relevant difference between what biotechnological enhancement does to human nature and what nature has done through evolutionary and environmental

---

[19] Whether there is such a duty or whether the recipients of efforts to enhance functions and traits in this way genuinely benefit from them is beyond my scope here. My point is simply that technologically induced divergences in functions or traits have not threatened the recognition of fellow humanity in these instances.

[20] It is important to stress that the recognition is "in principle." In practice, there is evidence that we have difficulty recognizing members of out-groups in ways that accord with our moral principles. Of course, this is also the case apart from biotechnological enhancement. But the latter could provide new occasions for in- and out-group identifications. Such an outcome would of course be problematic. But it seems likely that we would continue to recognize significantly enhanced persons in principle as fellow humans, just as we do now in analogous cases. I am indebted to Michelle Marvin for calling my attention to this matter and informing me of the relevance of research on intergroup empathy to it.

causes. Human enhancement would have to go to rather extreme lengths to justify the worry of some defenders of NS2 that human goods or rights are imperiled by biotechnological alteration of human nature.

We could let the first argument rest there. But Fukuyama might respond to it by insisting that even if the large-scale changes he worries about are not on the near horizon, it is likely that biotechnology will eventually be capable of introducing qualitative or much more significant quantitative changes to human functions or traits, to the point where it might be appropriate to speak of a change in human nature (or at least in the nature of some humans) or even of the emergence of a kind of being that is no longer human. At that point, his worry about the threat to liberal democratic rights and values would be relevant and, he might argue, we should therefore take regulatory measures now to make sure that we will never face such a situation. But why on Fukuyama's grounds should we assume that such a situation would be worse than our present one? If we grant Fukuyama's premise that liberal democratic values and human nature in its present form belong together, why should we assume that equally worthy or perhaps superior values would not come with a radically enhanced human nature? These questions point to the most fundamental flaw of Fukuyama's argument, which is its circularity. On the one hand, Fukuyama argues that liberal democratic values such as freedom, justice, and equality are more well founded than others because they relate to needs and interests that are deeply rooted in human nature. On the other hand, he argues that we should preserve human nature as it is because it is the condition of liberal democratic values. But even if it is true that liberal democratic values are more deeply rooted in human nature (in its present form) than other values, why should we commit ourselves to the state of affairs in which human nature and liberal democracy coincide? Why not allow biotechnological enhancement to bring about another state of affairs, one in which some other biological nature grounds some other (possibly superior) values? We may formulate the point of these questions in a more general way: If the basic claim of NS2 – that human goods or rights are grounded in human nature – is correct, then we must assume that different goods or rights would attach to a radically altered nature, and we could not reject radical enhancement, as Fukuyama does, simply because it would threaten the current nature-good or nature-rights relation. We would have to show how the current nature-good or nature-rights relation is superior to the prospective one and thus worthy of being preserved. Fukuyama has not attempted to do anything of this sort but has merely assumed the superiority of the present relation.

To conclude our consideration of the first argument, it appears that the variation and change that biotechnology is likely to introduce into human nature in the foreseeable future pose no threat to the grounding of human goods and rights in human nature. NS2's concept of human nature is flexible enough to accommodate a good deal of biotechnologically generated variation and change without undermining human nature as a stable ground of human goods and rights just as it has accommodated the variation and change that have resulted from the operation of evolutionary, environmental, and cultural forces. In neither case do variability and change rule out the grounding of human goods and rights in human nature.

What Matters: Our Good or Our Nature?

The last point made against Fukuyama brings us to the second argument. Some bioethicists argue that even if human goods or rights depend on human nature as NS2 holds, normative status properly attaches to those goods or rights, and not to the nature on which (at least for now) they depend, so that it is justifiable in principle to alter that nature if doing so would serve to promote or protect those goods or rights. If this argument were explicitly directed at NS2, it might go somewhat as follows. As long as we have no ability to alter our biological nature, we have reason to look to our nature in its current state to determine what counts as our good even if we can imagine, and find attractive, activities, capabilities, states, and conditions that our nature in its current state does not make available to us. These activities, and so forth, might involve the sorts of things proponents of biotechnological enhancement often commend: higher levels of cognitive capability; expanded or even entirely new perceptual capacities; a richer, narrower, or simply different range of emotions; greatly increased physical strength or agility; and a greatly extended life span. However, the argument continues, if we achieve the ability to intentionally alter our biological nature, it may no longer be necessary to look to our nature to determine what counts as our good. We may instead have the option of remaking our nature in accordance with goods such as those we just mentioned – goods that, even as we recognize them as (at least potentially) good for us, are unavailable to us in the current state of our nature. Under these circumstances (assuming they come to pass), it will no longer be sufficient for subscribers to NS2 to identify or commend something as good for us simply because it fulfills our nature or is more consonant with it than its alternatives are. Instead, they will have to decide whether normative status properly attaches (1) to human nature (in its current state) as the ground of the human good, in which case we must be willing to forgo some things we recognize as good and may be able to attain, or (2) to what we recognize as good, in which case we must be prepared to accept in principle the alteration of our biological functions and traits, perhaps even to a point where we may speak of a change to our nature or even a change of our nature into something else. If we decide in favor of our nature as it is, and thereby forfeit goods we could attain by its alteration, we have reason to question whether our nature truly is ordered to our flourishing. If we decide in favor of what we take to be our good, we no longer look to our nature to determine what our good is. Of course, the point of NS2 was to avoid a choice between our nature and our good. And with this forced choice, the two benefits of NS2 for Christian ethics – that it accounts for the goodness of creation and the intelligibility of the good, and that it provides a basis for the ethical evaluation of biotechnological enhancements – are lost.

Normative Status, Goods and Rights, Human Nature

So might the second argument go. As I have noted, the bioethicists who would press it have already decided for the second option. This brings us to the crux of the second argument. According to these bioethicists, even if goods are grounded in human nature in the ways that the various versions of NS2 suggest, normative status properly attaches to those goods and not to the nature on which they are grounded. We should value what is good, they argue, not what is natural. And in evaluating prospective biotechnological enhancements, we can and should consider them in abstraction from the limitations of our nature in its present form. If in our judgment they are genuine goods, yet our nature in its current state is not adequate to them, then it is in principle justifiable to alter our nature to make them available to us.

It is not surprising that self-professed transhumanists, who hold that humans in their current biological state are not all that they could be or are destined to become, welcome the posthuman prospect that this argument admits, but mainstream bioethicists can also be quite sanguine in the face of that prospect. John Harris argues with regard to potential population-level changes to human beings that "if the gains were … (sufficiently beneficial) and the risks acceptable, we would want to make the relevant alterations and [would] be justified in so doing, indeed … would have an obligation to make such changes." But that is all, morally speaking, that is at stake in the matter; "whether any proposed changes amount to changes in human nature, or to [*sic*] involve further evolution, seems ethically uninteresting."[21] Similarly, Allen Buchanan insists that "if the eventual cumulative result of a series of biomedical enhancements were to 'destroy' human nature by replacing us with beings that were 'posthuman,' that would not be a wrong in itself and might in fact be a good thing."[22] Jonathan Glover is more reserved, but his understated formulation puts the point all the more clearly and poignantly: "If a good argument showed that some terrible characteristic – which by genetic means we could change – was essential to being human, it might be better to transcend the limits of humanity rather than stay as we are. The idea of what is essential is a murky one, but, even if it were not, its importance is unclear. What is worth preserving is what is *valuable*, and the connection between the two [that is, between what is valuable and what is natural] is not obvious."[23]

To this second argument, at least three responses are available to subscribers to NS2. All three responses maintain even in the face of biotechnological enhancement that our good is what fulfills us as the kind of being we are, and that we therefore can and should continue to look to our nature to determine what our good is. The first response argues that even though we recognize as good for us certain traits or levels of functioning for which our nature in its current state is not suited, we recognize these goods by extrapolating from goods that are available to us in our current state. What we recognize as good, then, we recognize as good for us as beings of our kind, even though it will be necessary to change our nature to have access to these goods. The grounding of our good in our nature, which is essential to NS2, is therefore preserved, as is the principle that we look to our nature to determine our good. The second response argues that even if we concede that the goods that biotechnology promises are worthy of choice and perhaps even superior to those that our nature in its current state allows us to enjoy, it would be foolish to attempt to alter our nature in the pursuit of these goods. Normative status attaches to our nature as it now is, though not because of reasons that support NS1, but rather because our nature in its present form is the ground of the only

[21] John Harris, *Enhancing Evolution: The Ethical Case for Making Better People* (Princeton, NJ: Princeton University Press, 2007), p. 37. In the quoted material Harris is summarizing the argument of his previous book, *Wonderwoman and Superman: The Ethics of Human Biotechnology* (Oxford: Oxford University Press, 1992).

[22] Buchanan, *Beyond Humanity?*, p. 115.

[23] Jonathan Glover, *Choosing Children: Genes, Disability, and Design* (New York: Oxford University Press, 2006), p. 84 (emphasis in original).

good that can be prudentially pursued. The third response argues that the goods available to us in the present state of our nature are worthier of choice than whatever goods we stand to gain by altering or bypassing our nature. According to the last two of these responses, even if we do have an obligation to benefit others, as Buchanan, Glover, and Harris all say that we do, we will not genuinely benefit them by altering their biological nature. But all three responses conclude that subscribers to NS2 are justified in maintaining its basic claim that our good is what fulfills us as the kind of being we are and that we look to our nature to determine what our good is. If so, then from a Christian theological standpoint, the conviction that the goodness of creation consists in the connection of our good with our nature as created by God will be vindicated. The next three subsections take up these three responses.

The First Response

According to the first response, even if we recognize as good for us capabilities, states, and conditions – such as higher cognitive functioning, new forms of perception, different emotional ranges or intensities, greater physical strength or agility, and an increased life span – for which our nature in its present form does not equip us, we recognize these as good by extrapolating from what we know to be good for us given our nature as it now is. Although it is true in such cases that we alter our nature in accordance with what we take to be our good, insofar as we recognize it as our good by extrapolating from what fulfills our nature as it now is, we are still looking to our nature to determine what our good is.

To clarify this response, I begin with a question posed to biotechnological enhancement by Leon Kass. "If ... we can no longer look to our previously unalterable human nature for a standard or norm of what is good or better," Kass asks, "how will anyone know what constitutes an improvement?"[24] Kass is pointing to an epistemic condition of knowledge of our good: How can we recognize something as genuinely good (or bad) for us if we abstract from our nature as it is? Without our nature as a standard, claims about what is good seem to be entirely ungrounded. Against this worry, Glover and Buchanan point out that we already make evaluative judgments about human nature that presuppose a standard that is distinct from human nature.[25] Consider, for example, how we commonly

---

[24] Kass, *Life, Liberty and the Defense of Dignity*, p. 132.
[25] See Buchanan, *Beyond Humanity?*, pp. 136–38.

judge aspects of our biological nature to be bad (for example, susceptibilities to diseases), imperfect (for example, limitations of our cognitive or perceptual capacities), or ill-suited to the kind of life in society that most humans now live (for example, our impulses to out-group aggression). Glover and Buchanan claim that in judging aspects of human nature to be bad, imperfect, or ill-suited to our lives, we presuppose a standard of good or perfection that is external to human nature. And despite the appeal to an external standard, these judgments are, at least in many cases, well founded. If so, judgments about our good that do not take our nature as the standard do not seem to present the problem that worries Kass.

However, subscribers to NS2 would deny that they appeal in such cases to a standard that is external to our nature. From a Christian perspective, at least some of the disvalued features identified by Buchanan and Glover may be attributable to human nature as fallen rather than as created good by God. Such may be the case not only with disease susceptibilities but also with features, such as impulses to out-group aggression, that render our nature ill-suited to our current circumstances. It may even be the case for some imperfections.[26] If we recognize as good for us a state in which these disvalues are absent or moderated, it may be because we are appealing to unfallen nature as a standard to evaluate aspects of fallen nature, not because we are appealing to a standard that is external to nature. We should also keep in mind that NS2 involves a conception of human nature that accommodates variation and change, whether it is the result of evolutionary and environmental forces or of biotechnological activity. To project a form of human nature in which bad, imperfect, or unsuitable characteristics of human nature are absent or greatly mitigated and to judge it superior to the present form of human nature is, from this perspective, to appeal not to a standard that is external to human nature but to a standard of ideal or properly functioning nature. In short, the standard for judgments about bad, imperfect, or ill-suited features of human nature may be the standard of unfallen human nature or of ideal or properly functioning human nature. And if based on that standard we proceed to alter functions and traits to mitigate or eliminate those disvalues, our action is in

---

[26] From a Christian perspective, judgments about imperfections of human nature more plausibly refer to human nature prior to its eschatological transformation, not to its unfallen state, because the eschatological perfection of creation exceeds the original goodness of creation. Of course, to the extent that the eschatological perfection of human nature is continuous with human nature as created, such judgments do not, strictly speaking, appeal to a standard that is external to human nature. However, to the extent that the eschatological perfection of human nature is accomplished by God apart from creaturely agencies and processes, including those of biotechnology, it is an inappropriate standard for judging what is bad, imperfect, or ill-suited about our nature at present.

principle consistent with the twofold claim that our nature is the ground of our good and that we look to our nature to determine what our good is. It appears, then, that the second argument (that we must choose between our nature and our good, and that normative status properly attaches to our good, not to our nature) is mistaken, and that the first response has vindicated NS2 in the face of this argument.

This point is worth elaborating. To argue that our nature in its current form is not adequate to what we recognize as good is not to argue that our nature is not the standard of what we recognize as good. We recognize goods for which our nature in its present form is not adequate by extrapolating from goods that fulfill our nature in its present form. Those goods, which include capabilities and states involving, among many other things, cognition, perception, emotion, strength and agility, and longevity, are susceptible to quantitative and qualitative variation: We can imagine greater degrees (such as measurable increases in memory, concentration, or agility) and other versions (such as night vision or a different configuration of emotion) of them than those that are currently available to us. Based on this imaginative extrapolation we form the intention to alter our nature to make these greater degrees and other versions of goods available to us. But in none of this does our nature cease being the ground of our good or do we cease looking to our nature to determine what our good is.

There are, however, two limitations to this first response of NS2 to the second argument. One limitation involves the fallibility of extrapolation. There is a risk that the goods we arrive at by extrapolation, and for the sake of which we alter our nature, may turn out to be false or inferior goods, and we may regret the changes we made to our nature to make them available to us. Glover identifies the most serious objection to extrapolation from what we currently value by pointing out that the traits that we are likely to value today (or at least those that members of Glover's tribe are likely to value) differ from the patriotism and piety that Victorian-era enhancers might have hoped to design or select for.[27] We might also be in the grip of biases that could result in the colonization of the biological realm by the same ableist assumptions that have already colonized the social realm. The point is that if we depart from human nature as it now is as a standard for our good (that is, if we cease to identify the good as that which fulfills our nature in its current form), we seem to be at the mercy of historical and cultural relativity as well as outright bias, running a high risk that we would diminish rather than enhance the flourishing of our descendants

[27] Glover, *Choosing Children*, p. 98.

with the alleged goods we design them for today. The further our extrapolations take us from what is available to us in the present state of our nature, the greater the risk that this will occur. This is obviously a serious limitation, and one that could have damaging effects. However, we should keep in mind that all judgments of what fulfills human nature are fallible and reflect biases. Increased degrees of extrapolation do bring increased risks of morally problematic and even tragic judgments; they therefore call for increased scrutiny. But the same specter of fallibility and bias haunts all our judgments about our good, including judgments about what fulfills our nature in its present form.

The other limitation to this first response of NS2 to the second argument is a more serious one. How can we be sure that what we recognize as good will remain within the bounds of extrapolation from what is good for us as the kind of being we now are? If we can imagine goods that are quantitatively and qualitatively different from those that our nature in its present form makes available to us, why assume that our imagination will come to a stop at the point where an altogether different nature would be needed to experience them? And if we can imagine, and recognize as good, cognitive or perceptual capabilities, or emotional or physical states, that would require us to become a different kind of being, then are we not attaching normative status to what we recognize as good rather than to our nature as the ground of our good, looking to the good to determine what our nature will be rather than to our nature to determine what our good is? In short, does not the second argument defeat NS2 after all?[28]

To block this scenario, subscribers to NS2 have at least two options. One option is to deny that the changes we make to our nature would ever result in our becoming a different kind of being than we now are. In other words, there simply is no point where our extrapolations would break the boundaries of the kind of being we now are. To be sure, profound differences in cognitive capabilities or emotional states would place individuals at very different points on the spectrum of biological characteristics. But if we hold to the classical definition of "man" as a rational animal, we may be sure that as long as any two individuals at any two points on that spectrum, however distant from one another the points might be, combine some form of rationality with a biological nature, they will both be members of the human kind. And in that case, even though their biological

---

[28] In such a circumstance, the metaethical claim of NS2 would still hold insofar as the altered functions and traits would ground goods that fulfill *some* kind of being. But as a normative ethical position, in which we look to our nature to determine what our good is, NS2 would be defeated.

functions and traits as well as the goods those functions and traits make available to each of them differ significantly, it is only a matter of diverse goods that fulfill them as the kind of creature they both are. So might a subscriber to NS2 argue. But the argument begs the question at stake. That question is whether certain differences in the functions and traits of two rational animals are better accounted for by explaining them as possibilities of the same kind of being or as possibilities of two distinct kinds of being. That question is not one that can be decided *a priori*, by simply defining all rational animals as members of the same kind. And in any case, to insist that two individuals at very distant points on the spectrum of biological characteristics are members of the same kind would stretch credulity in cases in which the differences were such that they did not engage in meaningful communication, form significant social bonds, or reproduce with one another.[29]

The other option for defenders of NS2 is to attack the scenario from the opposite end by denying that we would ever recognize as good *for us* capabilities or states that would fulfill another kind of being than the kind we now are. On this view, we would never exceed the bounds of extrapolation because it is only after we have already become another kind of being that we would recognize what fulfills creatures of that kind as good *for us*. Anything we recognize as good for us will be something that fulfills us as the kind of being we now are. NS2 is therefore preserved: Even if one wants to attach normative status to what we recognize as good for us rather than to our nature as ground of our good, what we recognize as good for us will always be something that is (or that we take to be) good for a being of our kind. There are no grounds, then, for any scenario in which we would try to become a different kind of being than the kind we are to make available to us some capability, state, or condition that, prior to the alteration, we recognize as good for us.[30] This argument seems to preserve NS2, but it is challenged by the claims of some transhumanists, who express desires to experience goods that transcend what they, as the human beings they now are, can enjoy or even fully imagine.[31] It is as the humans they now are that these transhumanists desire to experience such goods, yet the goods

---

[29] Of course, it might well be plausible to hold that the personal identity of an individual would be preserved in the transition of that individual from one kind (or species) to another. If it is preserved, then we would be justified in saying that the goods of the person at each stage are goods of the same person. However, they would not be goods of the same kind of thing, and that is what is at stake in NS2.

[30] This option is articulated by Daniel Groll and Micah Lott, "Is There a Role for 'Human Nature' in Debates about Human Enhancement?," *Philosophy* 90 (2015): 623–51.

[31] See Nick Bostrom, "Transhumanist Values," *Review of Contemporary Philosophy* 4 (2005): 87–101.

they desire could be experienced (and indeed, adequately identified) only by another kind of being than the kind they now are. In this case, the relevant goods are clearly not recognized as good by extrapolation from goods that fulfill our nature in its present form. It appears that the second argument challenges NS2 once again as these transhumanists face a choice between what they consider to be good and human nature as the ground of the good and argue that normative status attaches to the good (which they hope to enjoy one day) rather than to human nature as the ground of the good.

But what if this transhumanist position in fact supports NS2? Nick Bostrom, a transhumanist who has reflected on this matter, concedes that the alterations of our nature that would make at least some such goods available to us would amount to a change in identity, such that the one who recognizes the goods in question as good *for her* would indeed be someone different from the one who now professes a desire to enjoy them.[32] If this difference is not only a difference in the identity of the person but also a difference in the kind of being the person is, this concession supports the second option for defenders of NS2: Only a being of a different kind than the one we are would recognize as good for it capacities or states that do not fulfill us as the kind of being we now are. Once again, NS2 appears to be vindicated: Whatever we recognize as good *for us* will be something that in our judgment fulfills us as the kind of being we are.

However, the matter is not so easily settled. A transhumanist may think that there are objective goods that, though we do not recognize them as good for us, we acknowledge to be superior, such that we consider it to be a gain to become a being that can enjoy them. This transhumanist agrees that our identity is inseparable from our existence as the kind of being we are, so that whatever we recognize as good *for us* will indeed be something that in our judgment fulfills us as the kind of being we are. Yet, as a being of our kind, she desires to enjoy goods that she cannot enjoy or even fully imagine without becoming a being of another kind. She does not, then, recognize these goods as good *for her*. But she nevertheless has (or at least thinks she has) a reason for pursuing these goods, whether for the sake of a successor identity to herself or for the sake of her descendants. To that extent, she recognizes them as good, even if they are not good *for her*. And to that extent, she once again attaches normative status to what she apprehends as good, and not to her nature as the ground of the good. Of course,

---

[32] See Bostrom, "Transhumanist Values," p. 95.

this transhumanist would strike many people as an odd character. If the good she recognizes is admittedly not good *for her*, what would motivate her to pursue it (and to alter her nature or that of her descendants to be capable of enjoying it)? But perhaps she is not so odd after all. To be dissatisfied with the goods that fulfill humans as the kind of being they are, to suspect that there are potentially greater goods than these, and to commit oneself to becoming or generating a kind of being that can recognize and enjoy those goods, whatever they may turn out to be, is not in any obvious way unreasonable.

At this point, a certain kind of traditional Christian version of NS2 has a clear advantage over secular versions. The belief that God has destined the human creature for eschatological perfection rules out the transhumanist's dissatisfaction, suspicion, and commitment. For Christians who hold this belief, the good we cannot yet enjoy or fully imagine involves the perfection of the kind of being we are, not that of a different kind of being. From such a perspective, there is no reason to become a different kind of being to become capable of enjoying goods other than those we will enjoy with the eschatological transformation of the kind of being we are. For such Christians, then, the claim of NS2 that what is good for us will always be something that fulfills us as the kind of being we are is sustained. But the motivation I have just described is not obviously unreasonable for those who lack this hope. For them, if there are goods that seem to be objectively superior, and they can be enjoyed only by becoming a different kind of being, it does not seem unreasonable to want to become that kind of being. And if it is not unreasonable, it is because we are justified in attaching normative status to what we consider to be good rather than to our nature as the ground of our good. The second and third responses attempt to vindicate NS2 in the face of this apparent failure of the first response to the challenge posed by the second argument.[33]

---

[33] One might wonder whether there is not a third option for defenders of NS2. This option would understand the relationship between our alterable nature and our extrapolated good as a dialectical one in which we extrapolate from goods that fulfill us in our present state, alter our nature in accordance with those goods, extrapolate once again from goods that fulfill us in our now altered state, alter our nature once again, and so on. This dialectical relationship between extrapolation and alteration has the advantage of offering a highly plausible model of the way in which biotechnological enhancement might well proceed. However, the open-endedness of both our nature and our good that it suggests is closer to the position we will present in the next chapter as NS3 than to NS2, for which the ordering of our nature to our good and the intelligibility of the good in relation to our nature seem to require more stable notions of the relationship of our nature to our good than this dialectical model offers. I will therefore not consider it as an option for defenders of NS2.

The Second Response

According to the argument we are considering, the prospect of biotech-nological enhancement forces a crisis in which subscribers to NS2 must decide whether normative status properly attaches to goods or to the human nature in which they are grounded – a crisis that those who are most likely to press the argument have already resolved in favor of the former. The second response of NS2 to this argument contends that even if biotechnology holds the promise of goods that appear to be superior to those that are now available to us, it would be foolish to attempt to alter our nature in pursuit of them. According to this response, the only goods that we can reliably attain are those that are grounded in our nature as it is. This being the case, there is no practical conflict between our nature and what we recognize as our good because we have reason to pursue only those goods that are grounded in our nature as it now is.

The most sophisticated version of this second response appeals to evo-lutionary biology, arguing that the alteration of human biological func-tions and traits could upset the finely tuned relationship between human nature and the human good that is the result of the workings of evolu-tionary processes over many eons. Gordon Graham nicely articulates this concern: "The human genome … is the outcome of many millions of years of selection and adaptation. This process has made the existing genome hugely well adapted to the human condition, the circumstances in which human beings must not merely survive but thrive… Now the ambition of re-fashioning this genome more effectively must rest upon the supposition that the accumulated results of a little less than 200 years of biological science will enable us to do better than indefinitely many years of evo-lution have done. What possible reason could we have to think this?"[34] Graham's point seems plausible when we consider scenarios like the one that many people fear will result from efforts to extend longevity. In this scenario, stem cell technology is capable of rebuilding vital organs such as hearts, lungs, livers, and kidneys so that deaths due to the failure of these organs are averted, yet progress on slowing or reversing neuro-degenerative conditions lags. People now live much longer – a goal of proponents of life-extension technologies – but the added years only prolong their state of cognitive decline. If this scenario or one like it in fact unfolds, those

---

[34] Gordon Graham, "Human Nature and the Human Condition," in *Future Perfect? God, Medicine and Human Identity*, edited by Celia Deane-Drummond and Peter Manley Scott (Edinburgh: T&T Clark, 2006), p. 42. For a similar argument, see Fukuyama, *Our Posthuman Future*, pp. 97–98.

who appeal to this argument from evolutionary biology would not be surprised that it is difficult to calibrate our biomedical advances in a way that benefits the whole organism. After all, the proportionality of the good of the human organism to its nature, for all its imperfection, took shape over many eons. It is foolish to think that we with our limited knowledge and in such a short time could improve on this record with a few alterations here and a few there.

Graham's point resonates when we think of biotechnological enhancements in analogy to options on a new car: as individual items or packages of items we simply add to the existing organism. But biotechnological enhancement could proceed as unintended anthropogenic change to human nature has always proceeded, namely, gradually and incrementally. Moreover, Graham's point appears to rest on a mistaken assumption about biological evolution. As anyone who has lived past the optimum age of reproduction might attest, evolutionary processes have not selected – or at least have not adequately selected – for the full range of goods that we consider to be constitutive of overall human well-being. Moreover, nearly the whole of human evolution occurred when humans lived under conditions that bear little resemblance to the conditions most humans live under today. To flourish in these conditions may require characteristics (for example, strong dispositions to out-group cooperation) for which our evolutionary past seems not to have adequately equipped us. In short, evolutionary biology offers us no reason to accept the results of evolutionary processes as an ideal or even an acceptable state of affairs so far as the human good as a whole is concerned. To be sure, natural selection has conferred on us a high degree of adaptability to our physical environment, so evolutionary biology counsels us to exercise caution in altering functions and traits in accordance with our judgments about what is good. It has also equipped us for a wide variety of social arrangements – a point we should keep in mind given the tendency of promoters of biotechnological enhancement to take the values and conditions of life in their own society, or in a narrow part thereof (for example, the university town), as the standard for the human good. And in any case, Graham is surely right that we have not learned enough in 200 years to be confident that our intentional interventions will indeed promote and not threaten our overall good. These prudential considerations should weigh heavily in the evaluation of any program of human enhancement. But it is still the case that over its many eons biological evolution has operated with insufficient regard for what is now our overall good, so that we have no reason to forgo biotechnological enhancement altogether on the assumption that biological evolution is a more reliable means to our overall good than bio-technology could ever be.[35]

The Third Response

We are left, then, with the third response of NS2 to the second argument. This response claims that the goods that fulfill our nature in its present state are more choice-worthy than any goods we stand to gain through the biotechnological alteration of our nature. Even if biotechnology makes alternative goods safely and effectively available to us, this response asserts, we should reject them as inferior to the goods that are to be enjoyed in our engagement with our biological nature as it now is. If the goods that fulfill our nature as it now is are indeed superior to any goods that we stand to gain by altering our nature, then NS2 is vindicated: If we want to determine what our good is, we should look to our nature. In formulating the third response in these terms I have in effect invoked Leon Kass, who has argued that the worthiest human desires as well as the deepest and most meaningful ideals and attachments to others emerge out of conscious struggle with the limitations, vulnerability, incompleteness, neediness, and dependence of biological life. Martha Nussbaum has made similar argu-ments and like Kass has focused on the goods that inhere in struggle with mortality and other limitations and vulnerabilities of finite human life.

Nussbaum nicely captures the key point: "Our finitude, and in particu-lar our mortality, which is a particularly central case of our finitude, and which conditions all our awareness of other limits, is a constitutive factor in all valuable things' having for us the value that in fact they have."[36] As this remark suggests, Nussbaum has in view the entirety of human needi-ness, vulnerability, and limitation but focuses on mortality. In an earlier essay she contrasts the deep understandings of the world, rich engagements with others, and perpetuation through lasting deeds that are available to those who must die, on the one hand, with the idle pursuits and petty rivalries that fill the lives of the immortal Greek gods, on the other hand.[37] This list of goods connected with mortality may come across as somewhat snobbish, but in a later work she draws on the contrast with the gods to

---

[35] See Buchanan, *Beyond Humanity?* pp. 181–93.
[36] Martha C. Nussbaum, *The Therapy of Desire: Theory and Practice in Hellenistic Ethics* (Princeton, NJ: Princeton University Press, 1994), p. 226.
[37] Martha C. Nussbaum, "Transcending Humanity," in *Love's Knowledge: Essays on Philosophy and Literature* (New York: Oxford University Press, 1990), pp. 365–91. See also Nussbaum, *The Therapy of Desire*, pp. 226–28.

show how friendship, erotic love, self-sacrifice for one's country, the care of parents for children, and commitment to justice are accessible only to beings who, in contrast to the gods (and, we might add, to some transhumanist ideals), are aware of bodily needs, vulnerabilities, and limitations that are inseparable from mortality.[38]

For Kass, human good depends on biological nature in the sense that basic biological conditions or characteristics that, as in the case of mortality, are not necessarily unique to the human species, take specifically human form in desires, longings, and bonds of affection that elevate mere biological necessity to something dignified and noble. Thus, he argues that insofar as the mortality that we share with other living beings becomes a matter of conscious awareness in us, it gives rise to the resolution to live one's life in a meaningful way, willingness to devote one's life to a cause even to the point of self-sacrifice, and longing for transcendence that is at least partially fulfilled in one's perpetuation through one's offspring.[39] This list of goods is more patriarchal than Nussbaum's later list (consider, for example, the difference between caring for one's children and finding transcendence in one's perpetuation through them). But there is considerable overlap in the two lists and in their central point, namely that virtues and values that seem integral to a truly worthwhile human life are intelligible only in their relation to beings who must consciously come to terms with their mortality and more generally with the neediness, vulnerability, and limitation that are inseparable from mortality.

There is, however, a notable difference between the position of Kass and that of Nussbaum, at least in her later work. Unlike Kass (and Bernard Williams, whose position she has explicitly in mind), Nussbaum does not argue that the life of mortality with its characteristic virtues and values is superior to an immortal life or that the latter would be of lesser or no value. Rather, she argues that it is simply not an option for finite creatures such as ourselves. Because we cannot get outside the perspective on value that our finite lives offer us, "there is not much point in asking whether a certain life seems good from the point of view of creatures that we have no chance of ever being, or rather creatures becoming identical to which we would no longer be ourselves."[40] If the contrast is between our mortal lives and lives that are, strictly speaking, immortal, Nussbaum's stance is persuasive. To

---

[38] Nussbaum, *The Therapy of Desire*, pp. 227f.

[39] Leon Kass, *Life, Liberty and the Defense of Dignity*, pp. 201–29, 265–68. See also President's Council on Bioethics, *Beyond Therapy: Biotechnology and the Pursuit of Happiness* (New York: HarperCollins, 2003), pp. 183–92.

[40] Nussbaum, *The Therapy of Desire*, p. 230.

desire an immortal life in place of our own is to forfeit the goods that are available to us in our form of life, which is the only kind of life we can live. But biotechnology complicates this contrast, putting before us in the form of radical life extension the prospect not of immortality but of indefinite extension of life. To the extent that this prospect is realistic, it is far from pointless – it is, rather, of great importance – to attempt to evaluate the virtues and values of a vastly extended life, which in many relevant respects approximates immortal life, with those of our present lives. I believe that Nussbaum's general argument regarding finitude has an important implication for this more complex case, but I will return to it in the context of a treatment of Kass's position, which (unlike Nussbaum's) considers mortality and the neediness, vulnerabilities, and limitations of our biological nature that accompany it in light of biotechnology.

Kass explicitly argues for the superiority of the goods related to our biological nature by showing how the specifically human form of biological life is lived in distinctively human desires, longings, and attachments to others that emerge in our ongoing struggle with biological necessity (which for him as for Nussbaum involves the limitations, vulnerability, neediness, and dependence that characterize the biological nature that we share with other forms of life). Kass insists that these desires, longings, and attachments are richer and worthier than those we stand to gain by eliminating or overriding biological necessity. Three points about Kass's position are especially salient in the context of this chapter. First, as noted previously, many of the human goods he emphasizes (namely, the desires, longings, and affective bonds that characterize the distinctively human form of biological life) depend on the mere presence in human nature of basic biological characteristics such as mortality and sexual reproduction that, as regards their mere incidence in humans (and acknowledging the incidence of impairment of sexual and reproductive functions in individual humans), are unchanging and invariable features of human nature. Because these goods depend on the mere presence of these characteristics, they are not embellished by altering the biological characteristics in which they are rooted. Resoluteness, willingness to sacrifice one's life, and transcendence through offspring depend on mortality simpliciter; hence, they cannot be enhanced by altering the life span. Similarly, the perpetuation and transcendence of self that for Kass occur in procreation depend on sexual reproduction simpliciter; they cannot be enhanced by altering sexual functions. In short, the goods that for Kass attach to sexual reproduction and mortality require only the bare presence of that function and trait.

These goods are therefore not embellished by greater or diminished by lesser quantitative variations in sexual function or life span.

It follows that there is no question here of a choice between human nature and goods that, though they are grounded in human nature, may be fully realizable only by altering human nature. The challenge the second argument issues to other versions of NS2 – namely, the choice between our nature and what we take to be our good that the prospect of biotechnological enhancement seems to force on us – does not arise here, as the relevant goods depend on the mere presence of the functions and traits that ground them and not on any state or condition of those traits that can be affected by biotechnology. At least in principle, Kass succeeds in linking human goods to human biological nature as it is, without either deducing those goods from our biological nature (they do not inhere in our biological nature but in the conscious engagement with our biological nature that for Kass constitutes the specifically human form of biological life) or loosening their connection to our biological nature in such a way that their full realization might require alteration of the latter.

The second salient point is that because for Kass the worthiest human goods are constituted in our conscious struggle with biological necessity, these goods are lost or diminished not at some point where biotechnology succeeds in altering our biological nature but at every point where it enables us to bypass engagement with our biological nature. In other words, biotechnology does not imperil human goods only, or even primarily, by changing human biological functions and traits – a prospect that is in any case quite remote when it comes to basic aspects of our biological nature such as sexual reproduction and mortality – but by making it possible for us to evade the engagements with these functions and traits in which the human good consists. A few of Kass's favored examples illustrate this point. First, if biotechnology eventually succeeds in making human reproductive cloning feasible in terms of safety, effectiveness, and cost (currently an unlikely prospect), it will presumably not be the case that sexual reproduction has been eliminated and is no longer a characteristic of human nature. It will rather be the case that at least some people will bypass it, choosing to reproduce by cloning instead. By so choosing, they will forfeit the attachments and affective bonds that according to Kass are rooted in generating offspring through sexual union (for example, partial transcendence of self in perpetuation through a genuine other) in favor of what he takes to be the more superficial meanings and bonds that cloning makes possible (for example, continuation of self in perpetuation through one who is a repetition of oneself).[41] Second, and similarly, the satisfaction of attaining a degree of self-improvement through the exercise of one's capacities is not forfeited only at the point where, say, a neurological intervention permanently alters one's capacity to recall a shameful deed one now wants to be clean of. Rather, it is already forfeited when, in place of coming to terms with the deed and its effects on oneself and others, one puts oneself on a drug regimen that has only a temporary effect on one's brain chemistry and must be repeated to have the desired result.[42] Finally, to return to the example with which we began, the loss of the meanings related to conscious struggle with mortality (namely, resoluteness, readiness to sacrifice oneself for a cause, and partial transcendence through one's posterity) does not occur when one becomes immortal or can expect to go on living indefinitely but has already begun when one's desires and aspirations are shaped by the prospect of technological extension of the life span. We forfeit these meanings that depend on our struggle with mortality the moment we substitute a commitment to the project of radical life extension for the struggle with mortality and thereby form our desires and yearnings around the prospect of an indefinitely long life rather than around a mortal one.

This point is worth stressing in response to an argument that is often made in criticism of claims that biotechnology threatens goods that are related to mortality. A radically extended life, critics argue, is after all not an immortal one. Death still comes eventually to everyone, and the prospect of an untimely death due to an accident or violence is unchanged. It follows that whatever virtues, values, and meanings are associated with mortality will still be accessible in a vastly longer life, even if they may be somewhat diminished (just as they may already have become diminished in our case, relative to that of our shorter-lived ancestors). This argument, however, misses the point. For Kass and Nussbaum, the goods related to mortality depend not only on the fact of our mortality (which of course won't go away) but also on our willingness to endorse the moral posture mortality offers us and inculcate the virtues, values, and meanings that are grounded in it. The project of radical life-extension offers a prospect Nussbaum does not consider, namely, that someone might not only desire the virtues and values of (quasi-)immortal life but might commit to a

---

[41] See Kass, *Life, Liberty and the Defense of Dignity*, pp. 150–62.
[42] Leon Kass, "Biotechnology and Our Human Future: Some General Reflections," in *Biotechnology: Our Future as Human Beings and Citizens*, edited by Sean D. Sutton (Albany: State University of New York Press, 2009), pp. 19–22.

life-extension regimen that inculcates those virtues and values in concrete practices. But while Nussbaum does not consider this prospect, she has the antidote to it, namely, recognition that the virtues and values of mortality are not merely given with our biological mortality but must be reflectively endorsed and deliberately inculcated in the face of an alternative posture. Forfeiture of these virtues and values does not require immortality. It requires only that we endorse and cultivate virtues and values that are associated with immortality, and we already do that when we commit ourselves to the project of life extension, even if we don't succeed in adding a day to our allotted span.[43]

As Kass's three examples (reproductive cloning, memory, and radical life-extension) demonstrate, it is not by the alteration or elimination of human biological traits and functions that biotechnology threatens human goods but by enabling us to disregard these traits and functions: to bypass or override them, or at least to divert our desires and aspirations away from the goods that inhere in the struggle with them and toward what Kass considers the more superficial goods to be gained by avoiding that struggle. For Kass, the most problematic sense in which biotechnology implicates our nature is not in altering but in bypassing it, and thereby enabling us to substitute inferior yet superficially attractive goods that depend on alienation from our biological nature for the superior and more profound goods that inhere in conscious, engaged struggle with it.

This last theme brings us to the third salient point about Kass's position, which is also the point that best illustrates his response to the second argument. It concerns the kind of goods that inhere in the conscious struggle with biological necessity and their superiority to the goods that biotechnology makes available by eliminating or bypassing biological necessity. For Kass, to bypass or override the struggle with our biological nature is to forfeit the worthiest desires, ideals, and attachments in favor of less noble (though by no means illegitimate) goods such as mental and physical

---

[43] It is important not to exaggerate this point. Although the goods that Kass connects with awareness of mortality are forfeited when we cease looking to mortality and begin looking to life extension for what we value, it is also true that the magnitude of the loss of these goods will correlate with actual increases in longevity. To this extent, it is an actual change in human nature, and not just a way of bypassing it, that threatens the good. At some point on the scale of life extension we will have ceased to relate to our offspring, with whom we have shared a century or more of adult life, as our children, and we will no longer feel the urge to accomplish what we can before we die. However, in a more fundamental sense, the goods inherent in the awareness of mortality are for Kass already sacrificed at the point where one's desires and aspirations are directed to increased longevity rather than to the desires, longings, and bonds that arise out of the struggle with mortality. One embraces the meanings and yearnings of the former rather than the latter and lives one's life in a way that is shaped by those meanings and yearnings even if one has no realistic prospect of living longer.

comfort, tranquility of mind, and long life.[44] A distinct kind of value inheres in the conscious struggle with biological necessity, namely, dignity, which Kass repeatedly and consistently expresses in the language of the sublime: biological necessity is "elevated" and "ennobled" in the struggle with it. Kass's ultimate argument for the superiority of goods grounded in our nature as it is to goods made accessible to us by biotechnology is thus a kind of aesthetic one. He seeks to get us to see the human life lived in the elevation of biological necessity as more attractive than the life we may hope to live by circumventing or attempting to overcome biological necessity by way of biotechnology. To return to the example of the human life span, he hopes that we will find the life of resoluteness, devotion, and transcendence through offspring that awareness of mortality affords us more noble than the prospect of an indefinitely extended life in which we are increasingly liberated from anxiety in the face of our impending death but simply pursue whatever desires we happen to have until we grow tired of them. To take another of the examples mentioned previously, he hopes that we will find our achievements and efforts at self-improvement worthier when we understand and experience them as the result of the active exercise of our capacities than when they are at least partially mediated by drugs (such as cognitive stimulants or selective memory inhibitors) that interrupt the relationship between our conscious activity and our achievement or self-improvement. My improved score on the placement test or my attainment of peace of mind in the face of my memory of a shameful deed lose meaning, he thinks, to the extent that they are not fully intelligible to me as something I have accomplished through the exercise of my own cognitive powers.

The most important feature of Kass's position for our purposes is his twofold claim that superior goods or meanings attach to certain necessities, vulnerabilities, and limitations of our nature as biological creatures, and that these goods or meanings may become inaccessible to us when we use technology to evade conscious struggle with necessity, vulnerability, and limitation. Claims very similar to these are central to the position I will propose in Chapter 5 as NS4. If they can be sustained, they resolve the crisis of NS2 by exposing the choice between human nature as it is and the human good as a false choice (because of the inferiority of the goods to be gained by altering or bypassing human nature as it is to those we may enjoy by engagement with human nature as it is) and ruling out in principle the prospect that the pursuit of the good will justify the alteration

---

[44] Kass, *Life, Liberty and the Defense of Dignity*, pp. 17f.

of human nature or its replacement by something else. However, Kass is a naturalist. He claims that the goods or meanings that attach to these vulnerabilities and limitations of our biological nature are fully intelligible within the horizon of biological life. That horizon is in no way a narrow one; we have noted that Kass understands it in an expansive sense to include not only biological characteristics but also the desires, yearnings, and attachments to others that inhere in the struggle with biological necessity through which humans live as the biological creatures they are. But its expansiveness notwithstanding, nature for Kass is an overwhelming totality: nothing transcends it, directs it, or qualifies it from without. Because this is the point on which NS4 disagrees with Kass, it is worth dwelling on it.

The totalizing character of biological nature is perhaps most obvious in Kass's suspicion of reason. For Kass, as we have seen, biological nature is "lived nature"; it involves everyday life in its concreteness and in all the dimensions in which we live it: physical, psychical, social, and spiritual. It is "life in its immediacy, vividness, and rootedness."[45] Along with this appeal to the ordinary, Kass also describes biological nature as "elusive" and in its higher forms "irreducibly mysterious."[46] These two characteristics of biological nature – its everyday familiarity and its elusiveness – lead Kass to credit custom, taboo, and quasi-instinctual reactions to new technologies while opposing the Promethean presumption to lay bare or explain away the mystery of nature.[47] The ordinary and the mysterious seem to be epistemological opposites, one suggesting the transparency of the immediately familiar, the other what is ultimately opaque. Yet it is significant that both are resistant to reason. In sharp contrast to Porter and the important strand of the Thomistic tradition she represents, and to Nussbaum, for whom rational reflection is necessary to identify and inculcate nature-grounded goods, Kass minimizes the role of reason in directing natural inclinations to their fulfillment (just as he also differs from Porter and Nussbaum in minimizing the role of historical and social contexts in the specification of goods). For him, the distinctively human ennoblement of biological characteristics such as sexual reproduction and mortality is imbedded in institutions such as marriage that, in his view, can play their proper role only when their customary, prereflective status is undisturbed by rational examination or

---

[45] Ibid., pp. 134; see also p. 18.
[46] Ibid., pp. 93n., 296.
[47] See especially Kass, "The Wisdom of Repugnance," *The New Republic* 216 (June 2, 1997), pp. 17–26.

principles of justice (which Kass consistently associates with the abstractions of reason). And for Kass, just as the modern biological science that reduces biological nature to an object of disengaged rationality has undermined appreciation of the experience of biological nature in ordinary life, and with it our confidence in the desires, affective bonds, and forms of life rooted in this experience, so also the expansion of notions of liberty and rights beyond the narrowly political confines of early liberalism into areas once governed by very different norms of private life has eroded the social patterns, traditions of long standing, and institutions that have been the primary bearers of such meanings and bonds.[48] In other words, just as abstract reason erodes the desires, yearnings, and affective bonds that are rooted in the awareness of and struggle with natural necessity, so public norms of justice undermine the institutions that are the primary bearers of these meanings and ways of being. In all these respects Kass rejects every kind of critical rational or moral reflection on nature-grounded goods and their institutional expression, allowing only a kind of reflective awareness of our nature that itself remains within the horizon of biological life and is imbued with the immediacy, ordinariness, and mystery that characterize that life. Kass is rightly criticized for romanticizing biological necessity, for the patriarchal assumptions that lie behind his emphases on biological lineage and perpetuation through progeny, and for presuming that nature can be accessed directly, apart from all constructions of it. These are all serious problems, and in my view they are ultimately rooted in his totalizing naturalism, in which biological nature is greatly expanded, but only to absorb everything – even reflection on it – into it without remainder.

A less obvious but for my purposes crucial implication of Kass's totalization of biological nature concerns the meanings that for him attach to biological necessity. Once again, the antireductionist expansiveness of Kass's biological nature is at the same time a subtler reduction of something – in this case meaning – to biological nature. While for Kass the meanings that attach to our biological vulnerabilities, limitations, neediness, and dependence inhere not in those characteristics but in our conscious struggle with them, and while this engagement entails a partial transcendence of them – what Nussbaum calls "internal transcendence" to distinguish it from a condition, "external transcendence," that would remove us

---

[48] Kass, *Life, Liberty and the Defense of Dignity*, pp. 12–15, 44–45, 50–52, 135.

altogether from vulnerability and limitation[49] – these meanings have no reference to anything that transcends our biological nature, understood in Kass's expansive sense. Biological nature, albeit in a richly expanded sense, is thus made to bear the full weight of the human good. It is not clear, however, that it can bear that weight, and for this reason Kass's position (along with Nussbaum's) can be accused of engaging in adaptive preference formation, which occurs when we adjust our preferences to unfavorable circumstances, making a virtue of necessity by insisting on the superior goodness of what we cannot after all do anything about (as prisoners, for example, may come to prefer life in prison).[50] In the terms set by Kass's biological sublime, characteristics such as vulnerability, limitation, neediness, and dependence are ennobled by engaging them, while human life is diminished by eliminating or evading them. But it is difficult to deny that protest, resistance, and refusal in the face of these characteristics can also be ennobling.

From a Christian theological perspective these stances of protest, resistance, and refusal, and the desire for external transcendence more broadly, may, however ambiguously, attest the failure of our biological nature in our fallen state to reliably reflect God's purpose for it. They may also attest a more fundamental conviction that even in the prelapsarian creation the divine purpose for our biological nature is not inherent in our biological nature, however expansively the latter is understood, but in its relation to something that transcends it and determines from outside the point of its existence and characteristics. Neither an internal transcendence that would come to terms with the vulnerabilities and limitations of our biological nature nor an external transcendence that would free us of those vulnerabilities and limitations, but to be related in those vulnerabilities and limitations to God who transcends yet also embraces them – this is the position that I will elaborate with the help of Karl Barth and Kathryn Tanner under the heading of NS4. For now, however, it stands as my most fundamental objection to Kass's response to the second argument, which is so promising in its attachment of human goods to biological vulnerabilities, limitations, and so forth, yet so problematic in its subjection of all normativity – moral reason, social institutions, human goods – to the centripetal pull of biological nature.

---

[49] See Nussbaum, "Transcending Humanity."
[50] The classic analysis is found in Jon Elster, *Sour Grapes: Studies in the Subversion of Rationality* (Cambridge: Cambridge University Press, 1983).

Conclusion

According to NS2, normative status attaches to human nature as the ground of human goods or rights. From a Christian perspective, the grounding of goods and rights in human nature attests the goodness of creation, which can be seen in the intelligibility of goods and rights as ordered to our created nature and in the ordering of our nature to our good, and enables us to look to our nature to determine what our goods are rights are. Insofar as biotechnological enhancement technologies can alter or bypass human biological functions and traits, they pose the question of the viability and status of these nature-grounded goods and rights. This general question is posed through two particular ones. The first asks whether biotechnological alteration of human nature imperils nature-grounded goods and rights by disrupting the stability of human nature on which they depend. We answered that, just as NS2 accommodates the variability and change that human nature, thanks to evolutionary and environmental processes, currently exhibits without imperiling these goods and rights, so it can accommodate the variation and change that biotechnological enhancement introduces without imperiling them. NS2 thus leaves a wide though not unlimited scope for the biotechnological alteration of human nature. The second question, however, poses a fundamental challenge to NS2. Up to now, NS2 has been able to presuppose human nature in its present form as a constant and to determine on its basis what count as genuine human goods. But if biotechnological enhancement makes it possible to determine what human nature will be based on what we consider to be genuine goods, subscribers to NS2 face a dilemma. Do they go with the good and refit human nature to accommodate it, or do they hold to human nature as it is, even though it means settling for an inferior good? Either way, they seem to abandon NS2, which depends on a relation between the human good and human nature that both alternatives deny. And either way, the claim of NS2 regarding the goodness of creation is imperiled inasmuch our nature is found not to be adequate to our good, and our good is no longer intelligible as the fulfillment of our nature.

The three responses to this dilemma all try to salvage NS2 by arguing that only goods that are grounded in human nature as it is are reliably or genuinely good for us. The first response notes that we recognize capacities and states that are good for us by extrapolating from the goods that fulfill our nature as it now is. Even though it leads us to goods that would require the alteration of our nature as it now is, extrapolation preserves the fundamental principle of NS2 that what we recognize as good (and in this case seek to make attainable by altering our nature) is what fulfills us as the kind of being we now are. However, this response breaks down at the point where the goods we seek are good for a different kind of being than the kind we are. Christians who hope for an ultimate transformation of the kind of being we now are have no reason to seek such goods. Others, however, may have reasons, and the next two responses urge us to forego extrapolation altogether and to rest content with the goods that fulfill our nature as it currently is. The second response asserts that even if we can imagine genuine goods that require an altered nature for their pursuit, it would be unwise to change our nature to make them available to us because we will never be able to excel biological evolution in proportioning our nature to our good. We saw, however, that this response overestimates what biological evolution has bequeathed to us. Finally, the third response, exemplified by Kass and Nussbaum, argues for the superiority (at least for us as finite humans) of the goods that inhere in our struggle with our nature as it is to any of the goods to be enjoyed by evading or eliminating that struggle. We concluded that Kass and Nussbaum are right to find meaning and value in characteristics that involve vulnerability and limitation but wrong to suppose that this meaning and value are immanent to biological life, however broadly it is understood. Speaking theologically, to imagine goods that exceed the present capacity of our nature is not always or only a sinful refusal to be satisfied with what God offers us in our creaturely nature; it may express, even if in a distorted way, the insights that our nature in our fallen state does not reliably reflect the good for which we have been created and that the good for which we have been created is not fully intelligible in terms of our creaturely nature. If or when biotechnology makes available to us goods that we can now only imagine, it will therefore not be enough to refer us to our creaturely nature as it is and commend goods that are intelligible in its terms alone.

The last point suggests that the mistake of NS2 lies in supposing that the normative significance of human nature is found in human nature, or more precisely, in a good that fulfills that nature as it is, rather than, as NS4 will have it, in a good that is found in the role of that nature, still as it is, in our life with God. As NS4 will hold, the point of our biological life is found in its relation to something that transcends it (while also embracing it), grounding its meaning and worth from outside it. But before turning to NS4, there is another alternative to consider. According to this alternative, which we met in the context of the second argument, the goodness of our created nature lies in its capacity to be altered in accordance with our ideas of the good. The connection between our created nature and the good, which for NS2 attests the goodness of creation, rests for NS3 on the indeterminacy, open-endedness, and malleability of our created nature. This connection, however, is not already given with the created characteristics we now have but is realized in the ongoing alteration of our nature. We will now consider this third version of the claim that normative status attaches to human nature.

CHAPTER 4

Human Nature as Susceptible to Intervention

Most people who ascribe normative status to human nature in the context of biotechnology do so with the aim of limiting or blocking the determination of human biological functions and traits by biotechnology. This aim has been evident in appeals to NS1 and NS2, as we have seen. However, normative status can attach to human nature in ways that, at least implicitly if not always explicitly, justify the determination of human functions and traits by biotechnology. It is not difficult to see how such a justification might come about. Human nature is often described as essentially indeterminate, open-ended, or malleable. Sometimes these characteristics are ascribed to human nature in continuity with biological nature generally, so that they belong to biological life as such, while at other times they are reserved for human nature in distinction from the rest of living nature, which is thought to be more constrained by biological necessities than human nature is. But in either case, these characteristics are sometimes taken to be normatively significant. When normative status attaches to human nature in respect of these characteristics, it is only a short step to the justification of biotechnological enhancement, which may be regarded as in principle permissible or perhaps obligatory.[1]

I will refer to this view of the normative status of human nature as NS3. To a greater extent than NS1 and NS2, NS3 encompasses a disparate group of positions. For all their differences, a shared opposition to biotechnological determination of human nature unites those whom I have

---

[1] It is important to point out that not everyone who claims that human nature possesses these characteristics argues that biotechnological enhancement is justifiable. For one recent example in which an emphasis on the open-endedness of human nature is combined with arguments for genetic interventions to cure or eliminate diseases and against genetic enhancement see Paul Jersild, *The Nature of Our Humanity: A Christian Response to Evolution and Biotechnology* (Minneapolis: Fortress Press, 2009). However, the logic of Jersild's position pushes in the direction of approval of biotechnological enhancement in principle and the reasons he gives for principled opposition to the latter are unconvincing.

identified with NS1, while a shared adherence to a broadly Aristotelian connection of human goods and rights to human nature unites those whom I have identified with NS2. No comparable set of shared commitments can be found among those whom I identify with NS3. Nevertheless, debates over biotechnological enhancement include participants who find normative significance in the indeterminacy, open-endedness, or malleability of human nature and who appeal to that significance to justify biotechnological enhancement, while beyond the narrow circle of biotechnology debates authors who hold similar views about human nature and its normative significance formulate positions that could justify biotechnological enhancement. Prominent among those I include in these groups are Donna Haraway, Philip Hefner, Jürgen Mittelstrass, Ted Peters, James C. Peterson, Kathryn Tanner, and Laurie Zoloth, all of whom I refer to in what follows. If one were to take previous eras of debate over biotechnology and other transformative technologies into account, Pierre Teilhard de Chardin and (at least to some extent) Karl Rahner would also count as subscribers to NS3.[2]

Indeterminacy, Open-Endedness, and Malleability

To the extent that we can speak of NS3 as a coherent position, it is constituted by agreement that human nature is indeterminate, open-ended, and/or malleable, and that normative status attaches to human nature in respect of one or more of these characteristics. The terms *indeterminacy*, *open-endedness*, and *malleability* are my own, but they identify characteristics of human nature (or of biological nature more broadly, within which human nature is taken to be included) which all the authors I consider to be representatives of NS3 find normatively significant. By *indeterminacy* I mean that biological functions and traits as they are exhibited in individual human beings do not rigidly or perfectly conform to concepts of human nature, such as Jean Porter's forms or Francis Fukuyama's statistical norms, which were discussed in Chapter 3, and that such concepts therefore do not exhaustively account for human nature either in whole or in

---

[2] See Pierre Teilhard de Chardin, *Activation of Energy*, translated by René Hague (London: Collins, 1970); idem, *The Future of Man*, translated by Norman Denny (New York: Doubleday, 2004); idem, *The Phenomenon of Man*, translated by Bernard Wall (London: Collins, 1965); idem, *Toward the Future*, translated by René Hague (London: Collins, 1975); and Karl Rahner, "The Experiment with Man: Theological Observations on Man's Self-Manipulation," in *Theological Investigations, Vol. 9: Writings of 1965–1967*, translated by Graham Harrison (New York: Herder and Herder, 1972), pp. 205–24.

part. By *open-endedness* I mean that while forms, norms, or biological laws may place constraints on the development of human biological functions and traits, they do not, either in the case of the individual or that of the species, prescribe a rigid program for the actual expression of those functions or traits in organisms or impose a limit on what those functions and traits might become. Finally, by *malleability* I mean that human biological functions and traits are in principle susceptible of alteration by intentional human activity, including biotechnology.

The minimum threshold of the plausibility of any view of the normative status of human nature is a low one: As long as human nature possesses the characteristic(s) to which normative status attaches, it meets it. NS3 has no difficulty meeting this threshold. The partial malleability of many biological functions and traits is well established, and the variability and change that are endemic to biological processes make it impossible to exclude indeterminacy and open-endedness, respectively, from accounts of human nature. It is no surprise, then, that subscribers to NS1 and NS2 allow for indeterminacy, open-endedness, and malleability in their accounts. As we saw in Chapter 2, prominent versions of NS1 confer immunity from intervention on human nature in full recognition of (indeed as a response to) the malleability that biotechnology has disclosed (or perhaps introduced) as a feature of human nature. Subscribers to NS1 may (and in the cases of Habermas and Sandel do) agree with subscribers to NS3 that human nature is in principle malleable while disagreeing over whether its malleability can be legitimately exploited or not. For its part, NS2 acknowledges the indeterminacy and open-endedness of human nature that are emphasized by NS3. The metaphysical forms and biostatistical norms described by Jean Porter and Francis Fukuyama, respectively, can accommodate a considerable degree of indeterminacy and open-endedness, as Porter and Fukuyama both point out. For example, Porter's metaphysical forms are indeterminate insofar as, on her account, particular defining characteristics of a species are neither necessary nor sufficient conditions for membership in that species (that is, membership of an individual organism in a species is neither forfeited for lack of any particular characteristic nor guaranteed by possession of any particular characteristic), while Fukuyama's biostatistical norms are open-ended insofar as they allow for ever-shifting median points for human characteristics (that is, the norm for traits such as longevity or height may keep advancing forward indefinitely). In both these cases indeterminacy and open-endedness reflect not only the inevitable limitations of any theoretical model in exhaustively accounting for its objects, but also the complexity and flux of biological life.

In short, the simple claim that indeterminacy, open-endedness, and malleability are characteristics of human nature or biological nature generally is uncontroversial, and NS3 easily meets the minimum threshold. However, the relative plausibility of any view of the normative status of human nature in relation to its rivals increases to the extent that the characteristics it picks out as normatively significant can be shown to be more fundamental to or characteristic of human nature than those picked out by its rivals, and what distinguishes NS3 is its claim (whether explicit or implicit) that indeterminacy, open-endedness, and malleability are more basic to human nature than the Aristotelian or neo-Aristotelian views that underlie NS2 take them to be. In this section, I briefly describe four influential schools of thought that have argued that biological nature in general or human biological nature in particular is essentially indeterminate, open-ended, or malleable. If their claims (or others like them) are defensible, these schools of thought would (other things equal) support the relative plausibility of NS3 over NS2. An adequate examination of these schools and their claims is well beyond the scope of this study, but it is worth considering them briefly because they elaborate in detail claims that are typically made only at a highly general level by nonscientists who subscribe to NS3.

The first school of thought is the one inaugurated by Georges Canguilhem. In his landmark study in the history of medicine, Canguilhem rejects the notion that we can understand biological life in terms of metaphysical forms or statistical norms.[3] Specifically, he denies that we can represent the normal state of the human organism as a statistical norm from which pathologies diverge and which medical treatment attempts to restore, arguing instead that biological life must be understood as itself a norm-generating process. Instead of attempting to understand the health and disease of organisms as (respectively) their conformity to and deviation from given physiological norms, then, we should understand organisms as generating their own norms. Health refers to states of the organism that are more conducive to its norm generation and disease to states that are less conducive to it, while the process of norm generation occurs in environments that may hinder or support the norms that are generated but that in any case condition their generation. From this perspective, health and disease are not related in terms of physiological norms (whether metaphysical or statistical) and divergence from them; rather, both are understood in terms of

---

[3] Georges Canguilhem, *The Normal and the Pathological*, translated by Carolyn R. Fawcett in collaboration with Robert S. Cohen (New York: Zone Books, 1991).

the norm-generating capacities of living things. An optimally healthy state is one in which an organism can generate norms that enable it to thrive in whatever environment it is in, while a diseased state is one in which the capacity of the organism to generate norms still operates but is greatly inhibited. The difference between the two states of the organism is a difference in its capacity to generate norms, not a difference between a norm and a divergence from it, and biostatistical norms simply reflect in statistical form the results of norm generation; they do not indicate anything about the normal state of an organism.

Canguilhem thus invites us to understand organisms as fundamentally indeterminate (that is, their functions and traits cannot be understood in terms of allegedly given physiological norms) and as open-ended (that is, they continually generate their own physiological norms under more and less favorable environmental and, for humans, social conditions). As a study in the history of medicine, his account focuses on physiology and pathology, but this focus is significant for our purposes insofar as it directly concerns medical science and practice and thus has relevance for biotechnological enhancement. Canguilhem does not draw explicit normative implications from his position, and it would be unfair to attribute to him any claim that normative status attaches to nature. However, he leaves no doubt that organic life, with its norm-generating character, is not normatively neutral: "[L]ife is not indifferent to the conditions in which it is possible ...; life is in fact a normative activity... Living beings prefer health to disease."[4] On these grounds a proponent of human biotechnological enhancement might draw some implications from Canguilhem's position that he does not draw. In particular, if (1) norms are generated, not given, and norm-generation is fundamental to life as such, and (2) if health is the capacity to set new norms, a capacity for which living things strive, then we might reasonably infer that far from threatening or violating human nature, the pursuit of enhanced biological states appears to be a characteristic expression of it. Of course, this inference would not be sufficient to justify biotechnological enhancement, but it would establish enhancement as a human activity that is in accord with human nature and indeed with biological nature generally.

Canguilhem has been criticized for the residual vitalism that comes across in statements like those quoted in the previous paragraph, where "life" is represented as an active force, and this tendency in his thought cut against the grain of the biological sciences while he was writing. This period

---

[4] Ibid., pp. 126f., 130f.

(roughly the mid- to late twentieth century) was marked by the triumph of molecular biology, which brings us to a second school of biological science that appears to support NS3. The notion that biological nature is fundamentally malleable or susceptible of manipulation was fostered by the conviction that biological processes can be fully comprehended in terms of physics and chemistry. While this conviction has a long pedigree, it prevailed over its competitors during the second half of the twentieth century, while the development of recombinant DNA techniques in the 1970s gave impetus to the conviction that living things were not only explicable in terms of physical and chemical processes but also susceptible of reconstitution through techniques such as gene splicing. In Tamar Sharon's words, what resulted is a view of biological nature as "an assemblage of discrete and transferable elements."[5]

In recent years, the conviction that living things are in principle fully comprehensible in terms of chemistry and physics has been complicated by approaches that, under the heading of "systems biology," attempt to understand the relationship between molecular-level interactions and higher-level functions of biological systems. These approaches treat characteristics of cells and organisms not as reducible to basic chemical and physical processes or as the products of linear intracellular processes, but rather as emergent properties that can be accounted for by mathematical and computational modeling of complex cellular networks. More broadly, the trends toward genomics, proteomics, and other "-omics" in molecular biology have led to more modest assessments of the malleability of biological nature. Nevertheless, the notion that biological nature is fundamentally malleable remains, thanks in part to the arrival of CRISPR/Cas9 techniques of gene editing and developments in the field of synthetic biology, which aims at redesigning natural biological systems and even the construction of novel biological systems by combining established techniques such as DNA sequencing and molecular cloning with newer techniques such as gene synthesis. It is premature to draw inferences about the malleability of highly complex biological systems from the relatively simple ones with which synthetic biology currently works, and it is yet unclear whether gene editing will succeed in influencing the kinds of functions and traits that are of interest for biotechnological enhancement. Nevertheless, synthetic biology and gene editing have given new impetus to the assumption

---

[5] Tamar Sharon, *Human Nature in an Age of Biotechnology: The Case for Mediated Posthumanism* (Dordrecht: Springer, 2014), p. 116.

that biological nature, at least at its most basic levels, is inherently malleable (and in the case of synthetic biology, even constructible).

The two schools of thought about biological nature we have considered (Canguilhem's and molecular biology) focus on characteristics of living things as such. The implications for human nature follow to the extent that human nature is taken to be comprehensible in terms of biological nature generally. However, the indeterminacy, open-endedness, and malleability of nature has also been conceived in a way that focuses more directly on human nature or even distinguishes human nature from the rest of nature. Once again it is worth looking briefly at two very different schools of thought that exhibit this tendency: one that has emerged in science and technology studies and the other in German-language philosophical anthropology.

The third school of thought focuses on human nature but mostly to undermine distinctions between humans and nonhumans. Donna Haraway has famously challenged assumptions that there are clear or fixed boundaries between humans and other animals and between humans and machines.[6] For our purposes it is the human-machine context that matters, and here Haraway's emphasis on the cyborg as a figure of human nature in postmodernity is accompanied by an important trend in the philosophy of technology.[7] The claim that there are no clear or fixed boundaries between humans and machines may be presented as a claim about human nature or as the claim that there is no such thing as human nature. In either version, it can be taken as something that has been true of humans all along (that is, that humans have always been coconstituted by technology) or as something that is true of humans only in very recent times (that is, it is only in recent decades, due to the development of biotechnology, information technology, and so on, that the boundary between humans and machines has blurred or dissolved and human nature has accordingly

---

[6] See Donna J. Haraway, *Simians, Cyborgs, and Women: The Reinvention of Nature* (New York: Routledge, 1991); and idem, *Modest_Witness @ Second_ Millenium. FemaleMan_Meets_Oncomouse: Feminism and Technoscience* (New York: Routledge, 1997). Haraway's position has recently been taken up in the field of Christian theology by Jeanine Thweatt-Bates, *Cyborg Selves: A Theological Anthropology of the Posthuman* (Aldershot: Ashgate, 2012).

[7] See especially Bernard Stiegler, *Technics and Time,* I: *The Fault of Epimetheus* (Stanford, CA: Stanford University Press, 1994); and Peter Kroes and Anthonie Meiers, eds., *The Empirical Turn in the Philosophy of Technology* (Amsterdam: Elsevier Science, 2001). Stiegler's "technogenesis," i.e., the coconstitution of humanity and technology, is taken up by a Christian theologian, Elaine Graham, in *Representations of the Post/Human: Monsters, Aliens and Others in Popular Culture* (New Brunswick, NJ: Rutgers University Press, 2002); and idem, "In Whose Image? Representations of Technology and the 'Ends' of Humanity," in Celia Deane-Drummond and Peter Manley Scott, eds., *Future Perfect? God, Medicine and Human Identity* (Edinburgh: T&T Clark, 2006), pp. 56–69.

changed or disappeared). In my view, the most plausible position is that humans coevolved with technology and that the coconstitution of human nature and technology is itself a feature of human nature (though not necessarily exclusive to human nature). That is, the lack of fixed boundaries between humans and machines is a fact about human nature, not a ground for the denial that there is such a thing as human nature, and an adequate account of human nature, from its origins until now, must include its coconstitution by technology. In any case, human-machine hybridity entails indeterminacy, open-endedness, and malleability. As is the case with the molecular school, the technological realization of cyborg humanity to date falls short of its theorization, but the phenomenon of human-machine hybridization (whether organic-mechanical or organic-digital) is sufficiently established to count as a feature of human nature to which normative status can attach.

The final line of thought radicalizes the common insight that the biological nature of humans, in a degree that contrasts with other species, grossly underdetermines their being, so that they must to a greater extent than other living things determine their own being. What makes this tradition of thought about human nature radical is its rejection (carried out to different degrees) of essentialism and teleology. One important version of this position is rooted in a school of modern German philosophical anthropology whose lineage has been traced by Jürgen Mittelstrass from its origin in the thought of Friedrich Nietzsche to its subsequent development by Max Scheler, Helmut Plessner, and Arnold Gehlen. Thinkers in this lineage stress the open-endedness of human nature. Man, in Nietzsche's terms, is "the not-yet-determined being," or in Scheler's terms, "the 'X that can behave in a world-open manner to an unlimited extent.'"[8] In his own contribution to this tradition of thought, Mittelstrass stresses the incompleteness, at both the phylogenetic and ontogenetic levels, of human biological development. "There is no 'natural' fate in the becoming of man, as an individual or as a species, that might be definitely determined by biological laws."[9] Mittelstrass takes it as axiomatic that evolutionary biology rules out a fixed human essence; human nature, we now know, "is subject to fundamental changes." With the rise of molecular biology, however, the basic malleability of human nature has also become clear to us. "That man

---

[8] Jürgen Mittelstrass, "Science and the Search for a New Anthropology," in *Rethinking Human Nature: A Multidisciplinary Approach*, edited by Malcolm Jeeves (Grand Rapids, MI: Eerdmans, 2011), pp. 62–63.

[9] Ibid., pp. 63f.

can intervene into these changes himself has only become clear in the light of the new biology – an ability to change his own genetic constitution and that of his progeny. In fact, the *conditio humana* itself is changing: in the sense that now even man's biological foundations are at his disposal."[10] It follows from this condition that "self-determination is not only the fate of the individual, but is also the fate of humanity itself; it belongs to the essence of humanity."[11]

While I have chosen these diverse schools of thought somewhat at random, they are far from marginal to the theoretical biology and philosophical anthropology of the last century. If their claims that indeterminacy, open-endedness, and malleability are fundamental to biological nature as such or to human nature more narrowly are credible claims, they would lend NS3 a plausibility that far exceeds the minimum threshold and would give it considerable advantages in a contest with NS2. However, some factors suggest that the claims are exaggerated or at least one-sided. In the case of malleability, the disappointing record of human gene transfer technology calls into question the extent of the malleability of human biological traits and functions. Whether the new gene editing techniques will fare better is not yet clear. Meanwhile, in addition to synthetic biology, strong claims regarding malleability are often made in the context of research involving neural-nano and neural-computer interfaces. It is not yet known whether this research will succeed in rendering higher-level brain functions as pliable as the predictions of their promoters promise. But for now, at least, Mittelstrass's claim that "man's biological foundations are at his disposal" is clearly an exaggeration. As for indeterminacy and open-endedness, Mittelstrass likewise exaggerates the lack of biological constraints on ontogenetic development, which for him are reducible to the trivial observations that "there is no adulthood before childhood, no reverse ageing, no Achilles who is young until he dies."[12]

For his part, Canguilhem might be accused of one-sidedness from an Aristotelian perspective such as Porter's. His argument against the norm-divergence model of pathology, if it is correct, rules out final causes, which in Porter's account rely on the notion of paradigmatic instantiations of functions and traits (which serve as norms) against which variation and development are measured. However, it does not necessarily rule out her formal causes. In other words, while his position is incompatible with a

---

[10] Ibid., p. 65.
[11] Ibid., p. 68.
[12] Ibid., p. 64.

kind of biological inquiry that accounts for traits and behaviors of organisms as approximations to and divergences from a paradigmatic norm (whether metaphysical or statistical), it might be formulated in a way that is compatible with a kind of biological inquiry that describes species and explains their traits and behaviors as characteristic of their kinds, so long as these descriptions and explanations are not taken to exhaust all the traits and behaviors exhibited by individual organisms (as Porter concedes).

As I noted at the outset, an adequate consideration of developments that appear to support NS3 would require much more attention to the authors and schools I have mentioned and to many others that I have neglected, taking us well beyond the scope of this study. However, what I have said is sufficient to support the claim that any plausible theory of human nature must account for both the order and continuity of biological nature emphasized by NS2 and the indeterminacy, open-endedness, and malleability emphasized by NS3 and conceded by NS1. As we have seen, the neo-Aristotelian philosophies of biology that typically underlie NS2 take metaphysical forms or statistical norms to be definitive of the biological nature of living things, even as these philosophies acknowledge indeterminacy and open-endedness to be endemic to all biological life. Meanwhile, it is only a tendency to exaggeration and one-sidedness that prevents the various perspectives that underlie the diverse versions of NS3 from conceding that biological nature unfolds in accordance with developmental patterns, exhibits structure and even finality (in the sense that living things progress toward full and flourishing states), and remains identifiably human in spite of the blurry boundaries with nonhumans, even as these perspectives take evolutionary or physical and chemical processes or hybridity to be more fundamental to and definitive of biological nature than any forms or norms that might account for the biological kinds that emerge in these processes, operate in accordance with them, or proliferate and recombine due to them.[13]

To conclude this section, a debate over the relative plausibility of the conceptions of biological nature or human nature that underlie NS2 and NS3 would turn less on disagreements over the characteristics each view ascribes to biological nature or human nature than on the philosophies and theologies of biological nature or of human nature and the normative

---

[13] To take a concrete example, for NS2 the fact of speciation does not belie the order and stability of the species that are in existence at any given time, while for NS3 the existence of relatively stable species does not belie the primacy of the dynamic processes of speciation through which they come into and pass out of existence.

implications of these philosophies and theologies. These implications determine, among the characteristics all parties could, at least in principle, agree are at least partially constitutive of biological nature, which ones are taken to be more explanatorily fundamental and to possess normative significance. Thus, Porter and Fukuyama are interested in showing how, in spite of its complexity and flux, human biological life exhibits the order and stability that are necessary conditions for identifying, promoting, and protecting enduring (if not fully determinate) goods and rights, and that for Porter attests the goodness of human biological life as God's creation, while for their part, subscribers to NS3 (as we will soon see) are interested in showing how, in spite of the laws and structures that constrain it, the open-endedness and malleability of human biological life exhibit its suitability for the human vocation to image God or bring an unfinished creation toward its perfection.

Objections and Replies

In attaching normative significance to indeterminacy, open-endedness, and malleability, however, NS3 appears to be vulnerable to two immediate objections. As was the case with NS1 and NS2, these objections are likely to keep people from adequately understanding NS3 or giving it the attention it deserves, so it is once again necessary, as it was in the cases of NS1 and NS2, to address them before critically examining some important versions of NS3. According to the first objection, it is not clear how human nature can have any more than a trivial kind of normative significance when that significance attaches to characteristics such as indeterminacy, open-endedness, or malleability. In contrast to NS1, human nature, understood in terms of these characteristics, seems incapable of constraining our actions in any meaningful way, and in contrast to NS2, it appears too unstable to bear any substantive or enduring meaning. The normative significance that NS3 ascribes to human nature thus seems to be reducible to the trivial point that human nature is open for biotechnological intervention, placing NS3 in direct opposition to NS1, for which intervention is precisely what is ruled out. But if this is the only normative payoff it can offer, NS3 differs little if at all from positions that appeal to the same characteristics to block or dismantle attempts to attach normative status to human nature. It is unclear, then, that there is any point in attaching normative status to human nature in the case of NS3. According to the second objection, if the justifiability of intervention is the only normative significance that can attach to human nature as indeterminate, open-ended, or malleable, then NS3 appears to derive normative significance directly from its descriptions of human nature. From the indeterminacy, open-endedness, or malleability of human nature, it seems, NS3 infers that intervention into human nature is justifiable. Unless this appearance is deceptive, NS3 is a rather crude instance of the naturalistic (or "is-ought") fallacy, which is committed when normative conclusions are derived from purely descriptive premises.

As was the case with the common objections to NS1 and NS2, so it is with NS3: Some versions of NS3 are vulnerable to these objections, but the more sophisticated versions avoid them. In the first place, for these sophisticated versions of NS3 the significance of the indeterminacy, open-endedness, or malleability of human nature is not reducible to the jus-tifiability of intervention. Rather, these characteristics are the locus of substantive theological or ethical meaning. For some prominent Christian subscribers to NS3, including Philip Hefner and Kathryn Tanner, the inde-terminacy, open-endedness, or malleability of human nature is integral to the claim that human beings bear or reflect the image of God.[14] In their view, human nature as created by God is indeterminate, open-ended, or malleable, and these characteristics – or more precisely, the implication that human nature cannot be conceptually or practically contained – comprise at least part of what it means for human beings to be made in the image of an infinite and incomprehensible God. Moreover, for some Christian and Jewish versions of NS3, including those of Hefner, James Peterson, and Laurie Zoloth, the justifiability in principle of biotechnological deter-mination of human biological functions and traits follows from the claim that God intentionally left creation incomplete or unfinished so that the task of its perfection would fall on human beings (as the Jewish version is more likely to go) or would involve cooperation with creaturely agency, including that of human beings (as the Christian version tends to argue). For these versions of NS3, indeterminacy, open-endedness, and malleabil-ity are just the characteristics we would expect human nature to have if the completion or perfection of God's creative work is part of the vocation to which God has called human beings.[15] In both of these cases (that is, humans as made in God's image and creation as incomplete or unfinished),

---

[14] See Philip Hefner, *Technology and Human Becoming* (Minneapolis: Fortress Press, 2003), pp. 73–88; Kathryn Tanner, *Christ the Key* (Cambridge: Cambridge University Press, 2010), pp. 52–57; idem, "Grace without Nature," in *Without Nature? A New Condition for Theology*, edited by David Albertson and Cabell King (New York: Fordham University Press, 2010), pp. 363–68.

[15] For Hefner and Peterson, this vocation is central to the sense in which humans bear the image of God.

the normative significance of human nature goes well beyond the trivial point that human nature is open for intervention to include substantive theological claims regarding the status of human beings and their vocation with respect to their created nature. Based on these claims, biotechnological determination of human nature can be thought to fall within a mandate, taking on the force of a broad duty to complete or perfect creation or to exemplify the *imago dei*. If some subscribers to NS1 worry that biotechnological determination of human nature manifests pride, in the form of usurpation of a role or prerogative of the Creator, subscribers to NS3 worry that to remain idle in the face of our powers to alter and control nature is to be slothful and to shrink back from our God-given role in creation.

Despite these considerations, however, the first objection to NS3 is not entirely misplaced. As we will see, even subscribers to NS3 who articulate the theological premises I have just referred to nevertheless have difficulty saying what follows from attaching normative significance to human nature as indeterminate, open-ended, or malleable, aside from a vague principle of change, progress, or improvement or a mere principle of liberty to determine what our nature shall be. This difficulty is hardly mysterious, as it is far from clear how human nature understood in terms of these three characteristics could support anything beyond these principles of change, and so forth – unless, as Kathryn Tanner argues, it is by virtue of indeterminacy, open-endedness, and malleability that human beings are related to a good or end that exceeds or transcends human nature (a claim that grounds one of the versions of NS4 that I will consider in Chapter 5). On the latter view, of which Tanner's work is an exemplary instance, the three characteristics NS3 takes as normatively significant are intelligible in terms of a telos of human life that constitutes its ultimate perfection and that these characteristics equip human beings to attain. But (as we will see) versions of NS3 that reject such a telos, or fail in their attempts to articulate it, tend to reduce perfection to "end-less" (that is, never-ending and lacking a telos) progress or banal improvement. In both cases indeterminacy, open-endedness, and malleability turn out to be ideal properties for the assimilation of human nature to late-capitalist production, which takes the form of undirected progress and thrives on the maximal adaptability of bodies.

In the second place, the argument for the justifiability of intervention does not, at least in sophisticated versions of NS3, proceed directly from purely descriptive claims about human nature to normative claims. It is worth noting in this context that the so-called naturalistic fallacy can be avoided in at least two ways. One way is to incorporate normative content into one's description of biological nature in general or human nature in particular. When Aristotelian defenders of NS2 argue that the good of a thing is its fulfillment as the kind of thing it is, they deny that we can provide an adequate description of a living thing without including in our description what would count as its fulfillment as the kind of thing it is, and thereby making, or at least implying, judgments about its good. Similarly, descriptions of human nature or biological nature that emphasize indeterminacy, open-endedness, and malleability may incorporate normative content. For example, we have seen that Canguilhem argues that a proper description of biological life includes a normative dimension insofar as biological life is norm generating, while Mittelstrass argues that because human biological life is inherently open-ended, an adequate description of human nature must include the freedom with which humans determine what their nature will concretely be. In both these cases an argument that humans should be free in principle to determine their nature (an argument made by Mittelstrass but not by Canguilhem) can be made in a way that avoids the naturalistic fallacy insofar as the normative content of its conclusion is also found in its premises.[16]

Another way to avoid the naturalistic fallacy is to connect the description of human nature to an independent normative claim that establishes the normative significance of the description. This way is evident in the claims about human beings made in God's image and the unfinished or incomplete creation that we have already encountered. In both these cases a theological claim (namely, the way in which humans image God or the divinely conferred vocation of humans in a creation that God has left incomplete or unfinished) picks out the aspect of human nature (namely, its indeterminacy, open-endedness, or malleability) to which normative status attaches. In short, in neither of these ways is there any direct inference from purely descriptive characteristics of human nature to the normative significance that attaches to human nature in respect of those characteristics. It follows that neither these versions of NS3 nor others that exhibit a similar line of argument commit the naturalistic fallacy.

---

[16] This is not to say that the premises supplied by Canguilhem and Mittelstrass are sufficient to justify the normative conclusions that might be drawn (and are in fact drawn by Mittelstrass) from their positions. In both cases something like an argument that it is good for something to act in accordance with its nature, or at least not to be prevented from acting in accordance with its nature, is required.

Human Nature in an Unfinished Creation

With these preliminary considerations behind us, I now turn to two theological versions of NS3: one that focuses on the creation as unfinished and thus in need of human action to bring it to its completion or perfection and another that focuses on the indeterminacy, open-endedness, or malleability of human nature as a constituent of the image of God in which humans are made and a condition of its full realization.

I begin with the notion of an unfinished or incomplete creation in Jewish and Christian approaches to biotechnological enhancement. Many Jewish bioethicists embrace the claim, articulated in an especially forceful way by Joseph Soloveitchik (whom we briefly met in Chapter 2) that God has left creation unfinished, thereby leaving scope for human action to transform nature.[17] "The Creator, as it were, impaired reality in order that mortal man could repair its flaws and perfect it." While Soloveitchik does not refer explicitly to the image of God in this context, the task of repairing and perfecting creation involves a kind of likeness to God. "Just as the Almighty constantly refined and improved the realm of existence during the six days of creation, so must man complete that creation and transform the domain of chaos and void into a perfect and beautiful reality."[18] Soloveitchik places this theme at the very center of Judaism. "The dream of creation is the central idea in the halakhic consciousness – the idea of the importance of man as a partner of the Almighty in the act of creation, man as creator of worlds."[19] It is important to understand that this point is primarily a practical one having to do with an ethical stance toward the world. Soloveitchik points out that for halakhic Judaism, the creation story in Genesis 1:1–2:3 is not myth or speculation but halakhah, namely, legal material, no less than are the *kedoshim* of Leviticus 19 or the *mishpatim* of Exodus 21. "If the Torah then chose to relate to man the tale of creation, we may clearly derive one law from this manner of procedure – viz., that man is obliged to engage in creation and the renewal of the cosmos."[20] The perfection of the world, then, is not merely an ideal; it is a religious obligation. It is indeed a mark of holiness. "If a man wishes to attain the rank

---

[17] Rabbi Joseph B. Soloveitchik, *Halakhic Man*, translated by Lawrence Kaplan (New York: The Jewish Publication Society, 1983 [1944]), p. 101.

[18] Ibid., p. 106. Soloveitchik refers most frequently to human creators as partners of God, but it is also clear that he thinks of their creative capacity as like that of God. See especially the section of the Talmudic tractate Sanhedrin quoted by Soloveitchik on p. 101.

[19] Ibid., p. 99.

[20] Ibid., p. 101.

of holiness, he must become a creator of worlds. If a man never creates, never brings into being anything new, anything original, then he cannot be holy unto his God."[21]

Soloveitchik does not explicitly mention technology, and the year of publication of his book (1944) predates the current era of human biotechnology. However, remarks such as those I have just quoted readily lend themselves to support for the biotechnological alteration of human nature. The notion that creation is unfinished coheres with the claim that human nature is open-ended; the divine enlistment of humans as partners in the work of creation coheres with the claim that human nature is susceptible of such action and is thus malleable; and the notion that God has, as it were, impaired the world leaves a wide scope for alteration of human nature. It is not surprising, then, that Jewish proponents of biotechnological enhancement have appealed to Soloveitchik in support of their position. Among them is Laurie Zoloth, who has eloquently argued in a series of publications that apart from considerations of justice and harm to others, halakhic Judaism imposes no restrictions on the alteration of human nature to cure or eliminate diseases or to enhance human functions and traits.[22]

It is widely but mistakenly assumed that Soloveitchik expresses a characteristically Jewish view of creation while the characteristically Christian view falls along the lines of the Augustinian convictions reflected in Oliver O'Donovan's version of NS1, which was examined in Chapter 2. As I noted there, this assumption ignores an important strand of Christian tradition that is commonly invoked in one form or another by Christian proponents of biotechnological enhancement. Just as some Christian subscribers to NS1, such as O'Donovan, appeal to Genesis 1:31–2:3 to declare creation a finished work, others may appeal to God's enlistment of creaturely agencies in the work of creation (Gen. 1:11, 20, 24), the blessings/commands to be fruitful and multiply (Gen. 1:22, 28), the command to the human creatures to subdue the earth (Gen. 1:28), and the vocation to till and keep the garden (Gen. 2:15) in support of the claim that creation is at least in some sense an ongoing project in which creatures participate.[23]

[21] Ibid., p. 108.

[22] See Laurie Zoloth, "The Duty to Heal an Unfinished World: Jewish Tradition and Genetic Research," *Dialog* 40 (2001): 299–300; idem, "The Ethics of the Eighth Day: Jewish Bioethics and Research on Human Embryonic Stem Cells," in *The Human Embryonic Stem Cell Debate: Science, Ethics, and Public Policy*, edited by Suzanne Holland, Karen Lebacqz, and Laurie Zoloth (Cambridge, MA: MIT Press, 2001), pp. 95–112; and idem, "Go and Tend the Earth: A Jewish View on an Enhanced World," *Journal of Law, Medicine and Ethics* (2008): 10–25.

[23] A clear statement of this general approach is found in Colin Gunton, *The Christian Faith: An Introduction to Christian Doctrine* (Oxford: Blackwell, 2002), pp. 3–15.

We should not assume, however, that these passages directly support the cause of biotechnological enhancement. While creatures play a direct role in creation according to the first set of verses, those verses that refer explicitly to humans can all be taken to mean that their role consists entirely in maintaining or keeping going what God has already created, and not in bringing into being novel kinds of things or even novel states or conditions of existing kinds of things. What is missing in these passages is the notion that creation in its original state is imperfect or "as it were, impaired," and thus lacking something that is to be attained by creaturely agency.

That notion, however, is precisely what is found in a well-known passage in Irenaeus' *Against Heresies*.[24] In this passage Irenaeus' characteristic insistence on the goodness of creation appears to be qualified by his admission that the first human pair lacked perfection. In God's creation of Adam and Eve, he asserts, the human being is not given all that it is destined to become but must rather attain perfection in time by passing through a succession of stages culminating in a likeness to God that is beyond the capability of human nature as such. While the attractiveness of this picture to proponents of biotechnological enhancement is obvious, we may doubt whether the passage in question lends itself to their cause. Is the perfection that Irenaeus has in mind commensurable with what biotechnology can deliver? Are the temporal stages he delineates identifiable with stages of technological progress? Is the overcoming of human nature by the love and power of God to which he refers something that is to be accomplished through the creaturely medium of biotechnology? We should hesitate before answering these questions in the affirmative, as it is clear from the passage that Irenaeus identifies perfection or likeness to God with immortality, which human nature is initially incapable of receiving and which is ultimately accomplished through Christ's sharing in mortal human nature and bringing it to its perfection. The distance from these soteriological claims to contemporary biotechnological enhancement is a long one, assuming that there is a viable route from the former to the latter at all.

If Irenaeus' position is to provide the justification sought by many Christian proponents of biotechnological enhancement, it will have to be distilled into the general claim that creation is not finished at the seventh day, as O'Donovan stresses in continuity with the Augustinian tradition,

---

[24] St. Irenaeus of Lyons, *Against Heresies* IV.38. In *Ante-Nicene Fathers*, Vol. 1: *The Apostolic Fathers, Justin Martyr, Irenaeus*, edited by Alexander Roberts and James Donaldson (Peabody, MA: Hendrickson Publishers, 1994 [1885]), pp. 521–22.

but rather continues through biological and historical time, over the course of which its perfection is attained. This distillation is just what one finds in some prominent Christian proponents of biotechnological enhancement. Ted Peters exhibits it in his proposal to relate *creatio ex nihilo* (the initial divine act of creation from nothing) and *creatio continua* (God's continuing work of creation) in the affirmation that to bring creation into existence is to give the world a future.[25] It appears also in Ronald Cole-Turner's suggestion that genetic engineering might be understood as a form of human participation in the ongoing divine work of creation.[26] Finally, it seems to be what James Peterson has in mind when, having cited numerous instances that illustrate the constancy of change in cosmological and terrestrial processes as initial grounds for his argument that human beings are called to improve nature (and human nature in particular), he sounds a distinctly yet also broadly Irenaean note: "The world looks as if the Creator … chose that creation would develop over time."[27] Of course, as we saw with O'Donovan, the completed status of creation may be taken to refer to its order while the things that are ordered are regarded as subject to temporal processes that are ultimately explained in terms of the finished order. But the claim that creation is not a finished work but one that continues in biological and historical time is parsed by at least some of these theologians into three additional claims that are near opposites of the Augustinian convictions we met in Chapter 2. First, *creatio ex nihilo* is interpreted not as an initial act of creation but rather as a reminder that creation depends on God as its source or that the cocreating activity carried out by humans is different in kind from the divine activity of creating.[28] Second, *creatio continua* is prioritized over *creatio ex nihilo*, while the significance of creation for human action and its theological meaning are found in the former rather than the latter. Creation is thereby historicized, so that dynamic evolutionary processes take priority over stable forms or norms and the origin of creation (however that is understood) is merely its starting point.[29] Third, the eschatological destiny of creation is emergent from

---

[25] Ted Peters, *GOD—The World's Future: Systematic Theology for a New Era* (Minneapolis: Fortress Press, 1992), pp. 129–46.

[26] Ronald Cole-Turner, *The New Genesis: Theology and the Genetic Revolution* (Louisville, KY: Westminster/John Knox Press, 1993).

[27] James C. Peterson, *Changing Human Nature: Ecology, Genes, and God* (Grand Rapids, MI: Eerdmans, 2010), p. 3. Irenaeus is directly cited and discussed by Peterson on pp. 6, 8f., 24f., 89, 127, 171.

[28] See Philip Hefner, *The Human Factor: Evolution, Culture, and Religion* (Minneapolis: Fortress Press, 1993), p. 43; and Ted Peters, *Playing God? Genetic Determinism and Human Freedom* (New York: Routledge, 1997), p. 14.

[29] See Hefner, *The Human Factor*, pp. 43–47.

intramundane reality, so that eschatological perfection is not deferred to the end of biological and historical time but is the culmination or perhaps merely the continual progression of immanent biohistorical processes.[30]

When one adds to these claims the thesis (which can also be supported by appeal to Irenaeus and, as we have seen, to Gen. 1) that God does not act alone in the work of continuing creation and eschatology but rather through the agency of creatures, and then goes on to grant technology a privileged role in that creaturely agency, one arrives at the position held by Philip Hefner, a theologian who is widely known for his proposal that we understand the human being as a "created cocreator," that is, "a creature who has been brought into existence by nature's processes, and who has been given by that nature the role of free co-creator within those same processes."[31] Human beings are both conditioned by natural processes and free with regard to them (albeit within them and not outside or above them); to put it in the somewhat oversimplifying terms he uses, they are a symbiosis of genes and culture.[32] Thus, Hefner's claim regarding the created cocreator is that culture emerges from biological processes, in the form of the evolved yet free human being, and thereupon becomes the instrument by which God brings these processes to their fulfillment. In short, while Hefner retains the notion of *creatio ex nihilo* as a reminder that the cocreator is a creature and not God, the substantive theological and ethical significance of creation for him consists entirely in *creatio continua* understood as ongoing natural processes that are formed by human culture (itself the product of those processes and always conditioned by them) in the direction of an eschatological destiny that is both immanent in and emergent from those processes as culture-formed yet fully natural.

The description of culture as God's instrument suggests that human freedom is not to be understood as the goal of the biological processes that produced it, and thus as the apex of creation (as an approach to biological evolution from a more traditional Christian perspective might well see it), but rather as a means to the comprehensive fulfillment of the whole of nature that Hefner identifies as God's ultimate purpose. Freedom is for the sake of nature, and not vice versa. "Indeed this project of taking the process of evolution … in new directions is what is at stake in the emergence of cultural evolution from biological processes."[33] As the title of his book

[30] See ibid., pp. 46–48.
[31] Ibid., p. 39.
[32] Ibid., p. 29.
[33] Ibid., p. 44.

(*The Human Factor*) suggests, Hefner's proposal is about the human role in bringing creation as a whole toward its eschatological destiny. More precisely, it is about how human freedom or culture serves the eschatological destiny of creation and thus carries out the divine purpose for creation. Christian faith, Hefner claims, understands the world "not simply in terms of what it has been and is now, but in terms of what it can become, and what it can become in light of God's intentions."[34]

No Christian theologian who (with Irenaeus and against Origen) thinks that the eschatological destiny of creation brings it to a perfection that goes beyond a mere return to the perfection of its original state would find the phrase I just quoted controversial. But Hefner understands it in such a way that *creatio continua* and eschatology bear all the normative significance of creation, thus denying that any theological or ethical meaning attaches to the original state. For him, the world "is fundamentally a realm of becoming," and this characteristic denies normative significance to God's initial creative act. "Although it is true that God had purposes for the creation from its beginning and that those purposes are in some sense programmed into the creation, the purposes of the creation cannot be extrapolated in any simple manner from the beginnings."[35] That formulation may appear to leave the door open to ascribing some normative significance to those beginnings as long as it is not done "in any simple manner," but Hefner appears to close that door when he goes on to reject "teleology," which "refers to preprogrammed goals that can be extrapolated from the original programming" of creation in favor of "teleonomy," which "speaks not of purposes that can be derived or predicted from natural structures, but rather speaks of them as consistent with these structures."[36] This formulation too is vague, and it might be interpreted in a way that is consistent with Porter or Fukuyama (that is, in terms of variation and change in consistency with formal causes or statistical norms). But whatever Hefner means by consistency with natural structures, he is clearly not interested in showing how any such structures constrain or limit what nature might become. For him, normative significance is found only in the open-endedness and malleability of nature and not at all in any feature of nature that might qualify these characteristics. Hefner is only adhering to his principle of open-endedness when he also suggests that our nature has no determinate eschatological destiny. " 'Human becoming'

[34] Ibid., p. 43.
[35] Ibid., p. 46.
[36] Ibid., p. 47.

expresses the idea that we are *always in process*, we are a becoming, and being human means that the journey is the reality – there may well be no *final* destination."[37] The conclusion is unavoidable: For Hefner, both our created goodness and its eschatological perfection are dissolved into pure becoming, and the latter is essentially a merely formal principle of continual change.

This last point (namely, the dissolution of creation and eschatology into becoming as a principle of continual change) and its implications are especially evident where Hefner focuses explicitly on technology. In his earlier work, Hefner understood the open-endedness of nature in terms of biological evolution, which in his quasi-Hegelian drama is propelled forward by generating culture, which in turn acts back on nature to transform it. For Hefner, the becoming of nature, as we have seen, occurs through the evolution of culture, but culture, as we have also seen, acts on nature not from outside but from inside, as that which is emergent from natural processes and remains within nature. Technology, moreover, is a crucial, and in our era a defining, aspect of culture. Thus, Hefner's position "places technology firmly within the evolutionary processes of the universe, recognizing it as a phase of cultural evolution."[38] As such, it plays an essential role in the becoming of nature, whether human or nonhuman. "Technology belongs to our becoming, it belongs to nature's becoming, and to the becoming of the universe. Technology is energy, the energy of an awakening and growing cosmos."[39] This theme, for which Hefner is deeply and explicitly indebted to Pierre Teilhard de Chardin, contrasts with the common tendency to think of and experience technology as alien from nature and of its practice as a violent imposition on nature. Hefner's first item of business when it comes to technology is to persuade us to substitute the view of technology as an expression of the becoming of nature (both human nature and the larger evolutionary and cosmic processes of which human nature is a part) for the view of technology as separate from nature.[40] But for our purposes what matters more is that Hefner proffers a view of technology as integral to and in part constitutive of creation. If creation simply is the becoming of nature – its progression, through cultural evolution, to its eschatological destiny (which destiny may be nothing other than the route, as we have seen) – and if technology is, in our era,

---

[37] Hefner, *Technology and Human Becoming*, p. 5 (italics in original).
[38] Hefner, *The Human Factor*, p. 49.
[39] Hefner, *Technology and Human Becoming*, p. 10.
[40] Ibid., pp. 1–27. See also Hefner, *The Human Factor*, p. 49.

one of the most definitive expressions of the cultural evolution through which nature becomes, then technology *is* creation: a paradigmatic and effectual instance of the progression of the continuing creation toward its emergent eschatological destiny. While Hefner does not refer to biotechnology in this context, it is clear how his position might justify biotechnological enhancement.

The centrality of technology to nature's becoming continues in Hefner's later work on technology, but the relationship of nature and technology is reversed, so that nature (including human nature) is now assimilated to technology. Echoing a prominent theme in the postmodern science and technology studies by Haraway and others that is briefly described earlier in this chapter, Hefner announces the breakdown of boundaries between nature, humanity, and technology, though he focuses on informatics, finding in its applicability to biological nature evidence that the latter is inherently open-ended and malleable.[41] Terms such as *self-generating, autocatalytic,* and *autopoeisis,* he points out, indicate a kind of freedom of unprogrammed behavior once thought exclusive to humans but now shared by humans, nature, and technology.[42] Thus understood, open-endedness and malleability possess ontological and theological significance. "If self-generation, autopoeisis – the making of ourselves – is written into the very substance of nature, as well as into the fundamental code of human nature and technology, we must consider that it is a clue to the nature of reality and, therefore, to the nature of God."[43] Nature is a restless drive for new possibilities, and "[w]hen we participate in this drive for new possibilities, we participate also in God."[44] Thus, while he speaks of the breakdown of boundaries between nature, humanity, and technology, what Hefner in fact does in this later work is assimilate nature and humanity to technology. In the end, nature and humanity turn out to be no more than autocatalysis and autopoiesis, that is, the endless, self-sustaining generation of novelty. But that is to say nature and humanity are the perfect image of technology, or more precisely, of techno-capitalism,

---

[41] Two important though very different studies that examine the emergence and implications of the view that humans and machines are intelligible in terms of information, which is conceptually and practically prior to both, are Jean-Pierre Dupuy, *On the Origins of Cognitive Science: The Mechanization of the Mind* (Cambridge, MA: MIT Press, 2009 [1994]); and Katherine Hayles, *How We Became Posthuman: Virtual Bodies in Cybernetics, Literature, and Informatics* (Chicago: University of Chicago Press, 1999).

[42] Hefner, *Technology and Human Becoming*, p. 50.

[43] Ibid., p. 81

[44] Ibid., p. 84.

which involves the endless production of novelty that is unconnected to considerations related to an antecedent need or good.

Like some promoters of artificial intelligence and unlike sophisticated theorists of technoscience such as Donna Haraway, Hefner forgets that when we portray human nature as an information system we are dealing with representations and not describing nature just as it is.[45] Insofar as he portrays biological nature as pure becoming that is only superficially differentiated and structured by processes that are characteristic of distinct kinds, he also uncritically adopts key features of the molecular view of biological nature and its reduction of the latter to chemistry and physics. But the most serious problems with Hefner's position are theological. First, already in his earlier work but especially in his later work, where autopoiesis becomes the key to ontology and theology, the becoming of nature is the true subject and ultimately the entire content of Hefner's theology. Whether becoming is understood primarily in terms of biological evolution, as his earlier work has it, or in terms of informatics, as in his later work, God and humanity are finally assimilated to becoming. It is doubtful that such a position can do justice to either the transcendence of God or the dignity of humanity as these notions are understood in mainstream Christian theology. Second, in the end, human nature for Hefner (and indeed, nature more generally) is simply the capacity to become something new. But if that is what human nature is, there can be no reason for God's having created us with the particular characteristics we now have. These characteristics exist only to be superseded; as such, they can have no normative significance. For Hefner, the goodness of creation is not ascribable to any state of creation as a whole or of its creatures (not even to a determinate eschatological end state) but only to the processes by which new states succeed old ones. Again, this view precludes the ascription of normative significance to any features that characterize human nature in the present (or at any other stage). Aside from the (general) property of indeterminate becoming, every (particular) property of the creation that God has brought into being is, from a normative standpoint, arbitrary, or at best merely provisional. With respect to human nature, aside from what is perhaps a greater capacity for becoming, there is, on Hefner's account, no reason why humans have the characteristics they do rather than some others. It is difficult to see how such a position can avoid compromising the Christian conviction of the goodness of creation.

---

[45] See Haraway, *Modest_Witness@Second_Millenium. FemaleMan_Meets_OncoMouse*.

A similar problem confronts James Peterson's effort to justify the genetic alteration of humans and creatures generally. Unlike Hefner, however, Peterson (as we will see) attempts to solve it by formulating criteria for the improvement of human nature by biotechnology. While Peterson stresses that particular judgments will always be necessary in concrete instances, these criteria present us with the general outlines of what it would mean for God's will for creation (and specifically for human nature) to be realized, or at least approximated, by biotechnological intervention. In principle, then, Peterson avoids the pointlessness of endless becoming. Peterson is explicitly Irenaean (and implicitly, recalls Soloveitchik) in his depiction of an unfinished and imperfect creation that humans are called to complete and perfect. That call is central to the image of God that humans reflect. "If we are to follow God, to reflect God's image, that includes extending God's creative action. As God develops us, we are to take part in developing his ongoing creation."[46] Peterson rejects the view that we should approach creation as an existing order that, as it merely happens to be, reflects the goodness of its Creator. "Even if one could clearly discern a natural order, why consider that a pure expression of God's will?"[47] One reason not to consider it such is sin, which "has warped our perceptions and tainted our physical world."[48] But another reason is that creation does not yet reflect God's will for it. "Accepting the premise that God created the natural world, one could still welcome it as a right and good starting point, not intended yet as a complete fulfillment."[49]

These convictions about nature are strongly determined by Peterson's awareness of the constancy of change, including anthropogenic change, and the pervasiveness of pain, disease, futility, and imperfection. As constantly changing and manifestly imperfect, nature (including human nature) is in its present state simply the sum of its characteristics and the relations among them that have evolved to date; it does not exhibit the perfection God wills for it, and to treat it as if it did is to slothfully disregard the vocation to which God has called us.[50] But in the absence of norms that provide a standard of that perfection, would not the same be true of technologically altered or enhanced nature? It seems that at any given stage, it too will be simply the sum of characteristics and relations among them that have evolved to that date (with their evolution now partly in the

[46] Peterson, *Changing Human Nature*, p. 23.
[47] Ibid., p. 29.
[48] Ibid.
[49] Ibid., p. 31.
[50] See especially ibid., pp. 10, 21, 83f.

hands of humans). This implication must be unsatisfactory to Peterson, whose version of NS3 repeatedly invokes the Irenaean conviction that creation is destined for perfection willed by God, and not merely for Hefner's perpetual becoming. Moreover, whatever is to count as perfection must be worthy of the name. As Peterson puts it, "This brief life on earth is a place of growth, a place of learning and becoming the kind of people that can glorify and enjoy God forever." He presses this point against biotechnological enhancements that would provide merely superficial pleasures and comforts. "The good life is not in amassing and securing personal comforts and prizes. It is about abiding joy, not just moments of happiness. It is about living in a way that pleases God and is suitable for the life to come."[51]

By gesturing in these ways toward a substantive vision of perfected humanity, Peterson seems poised to articulate criteria according to which biological enhancement might genuinely serve a recognizably Irenaean vision of human perfection and not merely continual becoming. However, notwithstanding a hint or two that point in this direction, the improvements to be sought through biotechnological enhancement amount in the end to the "primary goods" (including liberty, opportunity, wealth, health, intelligence, and imagination) identified by the philosopher John Rawls as rationally desirable by all persons, whatever their substantive conceptions of a good life may be, along with a similar set of goods ("mental acuity, mathematical and spatial reasoning, language facilities, creativity, musical abilities, and the like") identified by Ronald Lindsay, a bioethicist who has expanded Rawls's list.[52] It is unclear which of these goods Peterson thinks can be secured by biotechnological interventions, but it is worth noting that those having to do with health, intelligence, and creativity are frequently listed among the "all-purpose" or "basic" goods that (as we saw in Chapter 2 in connection with Habermas's claim that all substantive goods reflect merely subjective preferences) proponents of biotechnological enhancement consider to be conducive to any way of life a person might eventually choose and thus appropriate candidates for genetic choice by parents. In addition to these substantive goods, Peterson, like most other proponents of biotechnological enhancement, identifies safety, autonomy, and appropriate allocation of finite resources as side constraints on biotechnological interventions.[53] In sum, Peterson's criteria for biotechnological enhancement justify interventions when

[51] Ibid., p. 175.
[52] Ibid., pp. 176f.
[53] Ibid., pp. 163–204.

they aim at goods that promote any way of life one might choose and conform to basic bioethical norms. His proposal thus traces the familiar lines along which the standard bioethical justification of biotechnological enhancement runs. Irenaean perfection, it turns out, is indistinguishable from liberal eugenics. The problem with this conclusion is not that safety, autonomy, and appropriate use of resources are not important (they are), or that equipping people with biological characteristics that promote whichever way of life they might choose is a bad thing. But surely it is disappointing, having been enticed at the prospect of participating in the divine task of perfecting creation, to discover that the perfection is found in the realization of certain commendable but ultimately banal values of late bourgeois society.[54]

To summarize, for Peterson as for Hefner human nature is merely the sum of human characteristics and the relations among them that have evolved to date.[55] And in the absence of a norm of perfection the same will be true of any state of enhanced human nature: It will simply be what has evolved to date, now in part by intentional human action. Hefner and Peterson (especially the former) strike a progressive pose and hold out the promise that biotechnology will liberate us from our present limitations and from constraints our nature currently imposes on our well-being. It is worth pointing out, however, that what they offer largely coincides with the fundamental values of late modern capitalist societies. Hefner understands that his rejection of teleological and eschatological norms of perfection leaves him with mere becoming as his exclusive norm. I have pointed out the troubling implication of his position, namely, that no significance attaches to the particular created characteristics humans now possess, and that at no point in the past, present, or future does human nature reflect God's will for creation. At every point in the perpetual becoming, human nature as it is, is inadequate. Human life, like life more generally, is continual progress without a telos. This last point makes especially clear what

---

[54] To be fair, what Peterson envisions in *Changing Human Nature* is primarily the elimination of diseases and only secondarily the enhancement of traits. He looks to a future in which the genetic causes of misery and suffering are under human control and no longer threaten human beings and only secondarily to the prospect of improved human capacities. However, he denies that the distinction between treating diseases and enhancing functions or traits is normatively decisive. "The better question is how to enable the body to yield more support for human flourishing" (p. 125). And, rather than assigning disease to creation in its fallen state, he treats both disease and imperfection as characteristics of a creation that is not yet complete. With these moves he depicts both diseases and functions or traits in their imperfect states as constituents of a creation that does not yet reflect God's will for creation.

[55] Peterson, *Changing Human Nature*, pp. 10, 21, 81f.

has been the case all along, which is that Hefner has assimilated nature to techno-capitalism in its most basic form and has invested it with normative status on that ground.

In contrast to Hefner, Peterson recognizes the need for a substantive norm of perfection, in conformity to which an enhanced human nature would reflect God's will for creation. In principle, this move marks an advance over Hefner's position. In practice, however, what Peterson offers is nothing more than liberal eugenics with its "basic" or "all-purpose" goods. My point is not to impugn those goods, which are entirely appropriate in their proper context. But in a genuinely Irenaean vision that context would include an account of true human perfection. Without such an account, all-purpose goods do no more than suit people for the adaptability and productivity required by late modern economies. Whatever he might have set out to do, Peterson is unable to distinguish God's will for creation from the ideal of an omni-adaptive labor force.

The Alteration of Human Nature and the *Imago Dei*

The strongest theological argument for ascribing normative significance to human nature as indeterminate, open-ended, or malleable is found in Kathryn Tanner's account of the *imago dei*.[56] (A more thorough treatment of this theme appears in Chapter 5, where her conception is paired with Karl Barth's.) Tanner distinguishes two ways of understanding human beings in the image of God. One way attempts "to specify some set of clearly bounded, self-enclosed properties, a given human nature, which both reflects the divine nature and sets humans off from all other creatures."[57] Drawing on Greek patristic theologians such as Cyril of Alexandria and Gregory of Nyssa, Tanner proposes another way, one that begins with the Word of God, the second person of the Trinity, as the primary image of God, and only then turns to human beings, who image God in an important yet secondary sense. In Tanner's words, human beings "come to image God only when they take on that divine image and are deified, formed according to the divine Word as Christ's humanity was."[58] For Tanner these

---

[56] Tanner's explicit remarks on biotechnology are very few, brief, and mostly indirect, so to introduce her position in the present context is to risk exaggerating those remarks and making assumptions about their relevance to the present issue that she would not endorse. Nevertheless, her rigorous and thoughtful description of the plasticity of human nature and its theological significance justifies attending to her position in any account of the present view of the normative status of human nature.

[57] Tanner, "Grace without Nature," pp. 363f.

[58] Ibid., p. 364.

two ways of understanding the image of God correlate with distinct conceptions of human nature. The first way is associated with "an interest in a stable, fixed, and clearly demarcated human nature, the sort of nature that biogenetics calls into question."[59] The second way breaks sharply with any such conception of human nature. As Tanner puts it, "Since human beings do not image God in and of themselves but only when radically reworked into the divine image through Christ, it is the plasticity of human nature that becomes important. Plasticity is the capacity that allows humans to image God."[60] At least part of what Tanner means by plasticity involves what I have been referring to as *malleability*, namely, the susceptibility of human biological nature to intentional alteration. But plasticity as she understands it also involves indeterminacy and open-endedness. Because human nature is plastic it cannot be comprehended in determinate categories and what it may become is not strictly delimited by factors that have determined what it now is.

How is it that the image of God as Tanner describes it implies the plasticity of human nature? At the heart of Tanner's position is the notion of deification. For her, "God wants to give us the fullness of God's own life," which occurs as we receive the divine Word and are remade by it.[61] The question of human nature is posed as the question of the qualities or capacities human nature must have to receive the presence of the divine image and be remade by it. What must human nature be to receive and be transformed by what it is not, namely, the divine image? Tanner's answer is that humans have by nature an almost unlimited capacity to shape themselves by intentionally acting on themselves, and that it is this plasticity that equips them to receive the divine image and be remade by it. Meanwhile, this same plasticity resembles God's own unbounded and incomprehensible nature (here, the resemblance involves open-endedness and indeterminacy, respectively) and thus also images God in a weaker sense by what it is in and of itself.

Two important aspects or implications of Tanner's position are especially worth noting, partly because they enable Tanner to avoid the problems that beset Hefner and Peterson (along with other versions of NS3) and partly because they secure two of the most important claims of NS4, the position I develop in Chapter 5 (where I revisit Tanner's position). First, while Tanner's version of NS3 resembles Hefner's version insofar

---

[59] Ibid., p. 363.
[60] Ibid., p. 364.
[61] Tanner, *Christ the Key*, p. vii.

as it identifies the goodness of our created nature with its capacity to be other than it now is, it avoids the problems we have found with Hefner's version. We recall that Hefner's claim that our nature is simply the capacity to become something new left no room for any reason why God created us with the particular functions and traits we now have rather than some others, while his identification of our eschatological destiny with continual becoming (open-endedness) ruled out any telos for human nature. I suggested that these implications of his position left Hefner unable to adequately account for the goodness of creation. It is significant that Tanner avoids these problems, in part because the plasticity she attributes to human nature is not merely a principle of human nature as such but a feature of specific and identifiable human traits and capacities, and because the state of final perfection this plasticity serves consists in determinately Christian love.[62] Thus, she can in principle explain why we have the specific created characteristics we have (they are precisely the characteristics that equip us for our ultimate destiny, which is to be remade in the image of God), and she can describe what role these characteristics play in God's purpose for us. Meanwhile, although Tanner does not describe in detail the Christian love in which perfection consists, it provides her with the grounds, at least, for a richer conception of human perfection than the one we found in Peterson.

Second, it is crucial to emphasize that although the goodness of our nature consists for Tanner in its capacity to be other than it now is, our nature does not undergo any change of specific characteristics in receiving the divine image and being remade by it. Our creaturely nature is suited to the enjoyment of God's life by virtue of the characteristics with which God has equipped it. Put differently, our remaking in the divine image is possible due to the plasticity of the characteristics with which God has created us; it does not require us to acquire new or different characteristics (in contrast to theologies that hold that, for example, our sensuous nature is transcended and no longer operative in the state of blessedness). Because "it is the gift of God's own life" that we are given – that which we are not and never will be – our conformity to the divine image is not a gradual process by which our capacities incrementally approximate the divine image but rather a discontinuous leap in which "we are aided by grace to live in ways that are not natural to us because in keeping with divine power."[63] It follows that while our reception of the

---

[62] Ibid., pp. 40–50.
[63] Ibid., p. 62.

gift of the Holy Spirit brings about new created capabilities in us, such as the dispositions of faith and love, through which our lives are remade in conformity to the divine image, these new capabilities remain entirely dependent on the power of the Spirit to accomplish their end of conformity to the divine Word.[64] In conforming us to the divine image the power of the Holy Spirit does not insinuate itself, as it were, into natural human capacities so that the latter are supernaturally elevated yet still human. Instead this power retains its divine otherness, and we possess it only by clinging to what we are not.[65] Tanner's claim regarding the plasticity of human nature thus appears to amount to this: Human capacities which in themselves are incapable of conformity to the divine Word who is the image of God in the most proper sense are enabled by their plasticity – their open-ended, indeterminate, and relatively unbounded character – to receive the Holy Spirit as a power that is not in any sense their own and to be remade by this power to conform to the divine Word whose being and goodness are infinitely beyond anything they can claim as their own. Once again, then, Tanner can explain why we have the created characteristics we now have, and she can unambiguously affirm the goodness of those characteristics. It is with *this* nature, that is, with *these* specific characteristics and not some others, that we were created by God to enjoy life with God.

In both these respects, Tanner's position is exemplary for the position I will develop as NS4. Nevertheless, there is one point on which NS4 departs from her position and that justifies placing her position under the heading of NS3 rather than that of NS4. For Tanner, no normative significance attaches to our creaturely nature as such, apart from its being worked over by the Holy Spirit. She writes, "If God wants to give [God's own goodness] to humans, they have to be elevated beyond what they are themselves as creatures."[66] It follows that "Our nature is perfected and completed, ironically, by making us act unnaturally, in a divine rather than human way."[67] God wants to give us the fullness of God's own life, but to enjoy that life we must be taken beyond our nature as God initially created it, to the point that our subsequent condition can be described as unnatural with respect to our present condition and as divine rather than

[64] Ibid., pp. 83, 85
[65] Ibid., p. 104.
[66] Ibid., p. 60.
[67] Ibid., p. 62.

human. These claims amount to a strong conception of deification, and (to my mind, at least) they do not in the end do justice to the goodness of our nature in its present state and to the enjoyment of God's life that is meant for us in this state.

With its emphasis on the plasticity of human nature, its identification of the goodness of created human nature with its capacity to become other than it is, and its deificationist orientation, Tanner's version of NS3 may appear to supply a strong justification of the biotechnological enhancement of human functions and traits. However, it would be a mistake to assume that Tanner's position justifies biotechnological enhancement in principle. I will say more about this matter when I take up her position in Chapter 5, but for now it is important to emphasize that to claim that we must become something other than we now are to enjoy the fullness of life God wills to give us is not necessarily to claim that biotechnology is appropriate to the transformation that is required. As we have seen, Tanner emphasizes that the power to take us beyond our nature in its present state is the power of the Holy Spirit; it is not our own power or any power that we might summon or harness. The transformation of our nature through biotechnological enhancement may therefore amount to no more than a false semblance of the life God wants to give us and may even render us less suited to that life. Moreover, it is unclear from Tanner's position what biotechnological transformation could accomplish in the way of our deification. If our creaturely nature exists exclusively for a good that elevates it beyond itself, and if the characteristics of our nature as created by God are themselves suited to the transformation to which we are called, then we may wonder what rationale there is for improving creaturely nature. Nevertheless, Tanner's position does not rule out the possibility that at least some forms of biotechnological enhancement could turn out to be appropriate means by which the Holy Spirit transforms us into Christ's likeness.

Tanner does not address this matter, which she would presumably leave for Christian ethics to determine. For my purposes, however, what is most significant about her position is its contrast with the positions of Hefner and Peterson. Hefner, Peterson, and Tanner all agree that our created capacities have meaning or value only insofar as they enable us to become something other than what we now are. For Tanner, however, the divine image into which we are remade is not a mere formal principle of change, as it is for Hefner, nor does it cash out in the values of late bourgeois society, as it ultimately does for Peterson; rather, it has Christological content (even if Tanner says little about what exactly this content is). On its basis it is possible, at least in principle, to determine which forms of biotechnological enhancement, if any, would be worthy of pursuit (namely, those, if there are any, that facilitate closer approximations to the divine image).

In sum, Tanner affirms the indeterminacy, open-endedness, and malleability of human nature while also allowing for determinative normative content, and she shows how human nature has the characteristics it does because it is in or through these very characteristics that we participate in God's life. In these respects, her position points the way to the one I will present as NS4. At the same time, for Tanner as for Hefner our created capacities have meaning or value only insofar as they enable us to become something other than what we now are. While she can locate this meaning or value in specific human characteristics, and not merely in open-endedness and malleability as general characteristics, and can thus explain why God endowed humans with the specific characteristics we possess, she does not find meaning and value in the current state of human biological nature, and in that sense her position belongs to NS3 and differs from the position I develop in Chapter 5 as NS4.

Conclusion

In this chapter, I have considered the claim that normative status attaches to human nature as indeterminate, open-ended, and malleable. I have indicated what subscribers to NS3 mean by this claim, identified several accounts of human biological nature that might plausibly underwrite it, deflected two criticisms of it, and critically examined the major theological expositions of it. I have also discussed the extent to which this claim might appeal to Irenaeus as a precedent for it. Finally, however, I have found NS3 to be theologically unsatisfactory, and it is worthwhile to end the chapter with a concise statement of why I am not satisfied with it yet still believe that it has an indispensable place in the ethics of biotechnological enhancement.

One source of dissatisfaction involves the gap between the expectations NS3 engenders and the results that it is realistically capable of delivering. Proponents of NS3 typically present their position with the fanfare that often accompanies developments in biotechnology. There is something undeniably exhilarating in the discovery that human nature is susceptible of technological remaking and in the conviction that it is the human vocation to bring it to perfection by refashioning it. As is so often the case with developments in biotechnology, however, the results do not justify the hype, and the anticipated great strides forward for humanity turn out to look more like merely another stage in techno-capitalist progress.[68]

The major source of dissatisfaction with NS3 is, however, theological. Like NS1, NS3 is at bottom a story about creation and eschatology, and it is subject to criticisms of its approach to both topoi. First, neither the goodness of creation nor its status as a finished work is adequately defended by NS3. Creation for NS3 is a finished work in the sense that the creaturely agencies through which perfection will be attained are in place, and it is good in the sense that its indeterminacy, open-endedness, and malleability render it perfectible through the activity of these creaturely agencies in cooperation with divine agency. But on this view, the goodness of creation insofar as it is realized at any moment consists in its capacity to be other than it is on its way to perfection. It is therefore identifiable with the general features of indeterminacy, open-endedness, and malleability, but not with any specific features, such as those that characterize humans as the kind of thing they are. Humans and their kind (like all things and their

---

[68] Although this chapter has focused mostly on Christian theologians whose normative vision involves the perfection of human nature, this point also pertains to some secular thinkers whose normative visions are far less ambitious. Donna Haraway is a case in point. The figure of the cyborg that she and others have theorized represents a refusal of fixed boundaries and the identities, exclusions, and forms of power that depend on them. The crucial claim is that the notion of a stable human nature that is clearly demarcated from the nonhuman not only establishes absolute boundaries between the human and nonhuman, but its self-other binary also reinforces hierarchies within the human along the lines of race, sex, and gender. But with the breakdown of boundaries between the human and nonhuman in genetic technologies, organic-digital interfaces, and other developments, the pure, stable human dissolves into the ambiguity, heterogeneity, and alterability represented by the cyborg. The nearly endless possibilities of modification and reconfiguration of human nature that the cyborg represents undermine all descriptions of a stable human nature that could be privileged as normative and instead allow multiple and novel identities to emerge, thereby forming the conditions for a truly inclusive, antihegemonic politics. However, as Tamar Sharon insightfully points out in *Human Nature in an Age of Biotechnology*, the binary oppositions that Haraway's cyborg resists and disrupts are forms of power that according to her own analysis now belong to the past. The cyborg, she points out, "is the product of an age that is characterized by the *breakdown* of these same oppositions and dualisms" (p. 161). It is therefore unclear how it can oppose the forms of power that have, in her view, come to replace those of the recent past. As Sharon puts it, Haraway's cyborg "is simultaneously a *symptom* of postmodern conditions and an *agent of resistance* to postmodern power" (p. 162). But it is unclear how it can be the latter if it is the former. If Haraway is no more successful than Hefner and Peterson in avoiding the normative pull of contemporary techno-capitalism, it may be because the features of the cyborg she celebrates – its transgression of boundaries, disruption of stable hierarchies, proliferation of difference, and hybridity and fragmentation of identities – are celebrated just as enthusiastically in the biotechnology industry.

kinds) are at best a stage in an ongoing process leading to perfection, and the goodness of creation is not instantiated in their human characteristics as such, except as these characteristics enable them, at least for now, to exercise a superior form of agency in bringing creation to its next stage of perfection (at which point the agency of some other kind of being might well be superior). The ironic result is that NS3 does not really do justice to creation as an ongoing work either, despite its praiseworthy attempt to rescue this theme from tendencies to ignore or deny it, because it can only represent that characteristic of creation as endless or at least indefinite progress.

Second, NS3 presumes that the eschatological perfection of creation is commensurable with what biotechnology can accomplish – not necessarily that the eschaton will be realized in its fullness in biotechnological enhancement, but at least that biotechnological progress is capable in principle of approximating or tracking eschatological perfection. The problem with this presumption is its failure to see that any state of perfection that is pursued through biotechnological enhancement is subject to the limits to imagining "what no eye has seen, nor ear heard, nor the human heart conceived, what God has prepared for those who love him" (I Cor. 2:9). These limits are creaturely limits, and they mark eschatological perfection as something that God does not bring about along the lines of creaturely agency, even if God may enlist creaturely agency in anticipating or approximating it.

Despite these failings, NS3 makes a vital contribution to Christian ethics in the context of biotechnological enhancement. Its emphasis on the indeterminacy, open-endedness, and malleability of human nature yields two benefits that are difficult to derive from NS1 and NS2 (even if they are not in principle incompatible with NS1 and NS2). First, it radically undermines the racism, sexism, and ableism that thrive on constructions of human nature as fixed, unchanging, and intact. Second, it not only accommodates but positively affirms the ongoing creaturely agency that partly accounts for human nature in its present state and that is of course involved in intentional biotechnological determinations of human functions and traits. Finally, it offers the right answer to Elaine Graham's rhetorical question: "But does our concept of human nature have to be fixed and immutable in order to have moral substance?"[69] To find normative significance in the indeterminacy, open-endedness, and malleability of our

---

[69] Elaine Graham, "Bioethics after Posthumanism: Natural Law, Communicative Action and the Problem of Self-Design," *Ecotheology* 9 (2004): 193.

nature as created by God is to do justice to a line of thought in Christian reflection on creation that is not adequately represented by NS1 and NS2. It is NS3 that has brought this line of thought to bear on biotechnology, and on these grounds it deserves a permanent place in Christian approaches to biotechnology.

Our final word on NS3 must nevertheless be a word of warning. To attach normative status to human nature with respect to indeterminacy, open-endedness, and malleability alone is to imperil recognition of the goodness of our created nature as it now is.[70] And that is a problem that NS4 is well designed to correct, as we will now see.

---

[70] Tanner avoids this problem to the extent that her position is also an instance of NS4, as I will now go on to demonstrate.

CHAPTER 5

Human Nature as Condition for Imaging God

My examination of NS2 ended with the claim that from a Christian perspective the meaning or purpose of our biological nature is properly found in its relation to something that transcends it from outside while also embracing it, and not in an internally transcendent struggle with it. Our biological nature, even when it is understood broadly to encompass not only biological characteristics but forms of life that fulfill those characteristics (Porter) or ennoble the vulnerability and limitation they involve (Kass, Nussbaum), is not itself the ground of the goods that attach to it. To put it positively, our creaturely nature (of which our biological nature is a part) was made for life with God, and its meaning and purpose are ultimately intelligible only in that light. Taken by itself (that is, without reference to its ordering to life with God), our nature is incapable of grounding its meaning or purpose. From a Christian perspective, then, we have reason not to rest satisfied with goods that are limited to the horizon of our biological nature and not to endorse these goods as sufficient. This posture amounts to a rejection of what Nussbaum calls "internal transcendence," which is achieved in and through the vulnerabilities and limitations of our biological nature rather than in their overcoming. At the same time, however, this posture implies that the meaning or purpose of our biological nature *is* to be found in the relation of our nature as God created it to something that transcends it from outside. Specifically, it is to be found in the characteristics or capacities of our nature (including our biological nature) as our life with God is lived in them or through them, and not in a bio-technologically altered nature whose characteristics or capacities are different from those of our nature as it now is. We have reason, then, not to look to the biotechnological alteration of our nature as the solution to our dissatisfaction with the goods that our nature affords us and as the source of greater or more choice-worthy goods. This posture therefore endorses Nussbaum's rejection of "external transcendence," which would be achieved by leaving biological vulnerabilities and limitations such as mortality behind.

My examination of NS3 concluded with the claim that if our present creaturely characteristics and capacities merely amount to a stage on the way to something greater, then the goodness of our creaturely nature and the claim that creation is in some sense a finished work are compromised. We must be able in some way to identify our creaturely nature as we now know and experience it with the creation that Genesis 1:31–2:3 pronounces good and declares to be finished, even as we readily acknowledge that our nature in its present state came about through processes in time. Of course, this identification will have to be qualified as we account for the effects of sin on our nature and the eschatological fulfillment that awaits it, and it must also accommodate the biological facts that our nature undergoes continual change and will exist only temporarily (having existed for only a comparatively very brief time thus far). These qualifications and accommodation are significant, but they do not compromise or threaten the fundamental affirmation. One may consistently hold that our nature in its present form manifests the goodness and finished character of creation without holding that it is static or timeless, or that it already is all that it is destined to become in the eschaton. The crucial point is that if the claims that creation is good and is a finished work are true of our created nature in its present state, then they must pertain in some way to our present biological characteristics (or traits) and capacities (or powers). And this point implies in turn that the eschatological fulfillment of our nature, whatever form it takes, will involve the transformation of the nature we now have; it will not depend on the attainment through biotechnology of a nature we have yet to realize (indeed, it may even be incompatible with such a nature).

These convictions set important parameters for the position I now introduce as NS4. According to NS4, our creaturely nature suits or equips us for a particular form of life with God that was God's purpose in creating us. However they may have come about in time (and whatever is eventually to become of them in time, putting aside for now God's eschatological intervention to bring time to a close), we have the natural traits and powers we happen to have because it is these traits and powers, created by God, that constitute our *capacity* to enjoy the life with God for which God has determined us (that is, the traits and powers in, by, or through which we live our life with God) even as the actual *capability* to enjoy that life (that is, the ability or power to bring about and maintain our life with God) depends entirely on God's grace and is not given with our creaturely capacity. Normative status therefore attaches to human nature (which with its present traits and powers is to be understood as in some sense the good and finished work of God) as the condition for the particular form of life with God for which God has created us.[1] In this life with God, our nature relates us to God's own life, which in Jesus Christ both transcends us from outside, as the divine life in which we participate, and embraces us, as the human life that is lived with and for us. With respect to its status as God's creation, human nature is good insofar as it suits human beings for the life with God for which God has determined them, and it is a finished work insofar as it is the nature in or through which, by God's grace, they enjoy life with God. These are the defining claims of NS4.

Human Biological Nature and the Image of God

NS4 could be developed in several ways. One way, which has the formidable Thomist tradition on its side, would complete NS2 by articulating the supernatural end to which the natural goods that are grounded in our biological nature are ultimately directed and in light of which they must ultimately be understood. This way would remain within the parameters of NS4 insofar as its articulation of the supernatural end (that is, our life with God) relates our created nature to something that transcends it from outside while also affirming its characteristics and capacities in their created form. Insofar as a Thomist position treats our natural good as ordered to our supernatural end and incomplete apart from that ordering, it avoids the problem I have found with NS2 (that is, it does not limit the meaning or purpose of our biological nature to the horizon of our biological nature); and insofar as the supernatural additions to our natural capacities, which for Thomists are necessary for attainment of our supernatural end, extend those capacities and direct them to our supernatural end rather than replacing them with new capacities, it avoids the problem I have found with NS3 (that is, the attainment of our good does not depend on the emergence in time of capacities we do not yet possess). At the same time, certain tendencies in the Thomist tradition make it difficult to dissociate Thomism as clearly and decisively from the problems of NS2 and NS3 as I would like to do. For Thomists, our natural end is in principle intelligible apart from our supernatural end, even if it is ultimately ordered to the latter, while our supernatural end for its part is attained not in or

---

[1] This particular form of life with God, as we will see, is also a particular form of life with our fellow humans.

through our natural capacities in their created state but in their elevation and transcendence. In other words, the challenge of the Thomist tradition is to avoid treating human nature as normatively intelligible in itself apart from its relation to what transcends it from outside when considering the natural end of humans, and to avoid compromising the integrity of human nature with the supernatural transformations and circumventions of natural capacities it introduces into its consideration of the supernatural end of humans. Of course, to identify these challenges is not to say that they cannot be met; I suspect that they can. But in its characteristic forms, NS4 makes a strong claim about the sufficiency of our created nature – a claim that sets it off from NS3. I will accordingly focus on versions that avoid these challenges from the outset, leaving it to those who have independent reasons for adhering to Thomism to demonstrate how the latter can meet them.

The versions of NS4 that interest me treat human nature as a constituent of the image of God. The image of God is a difficult notion insofar as it seems to be both fundamental to theological anthropology and impossible to identify in a definitive way. The result is that it has been given many different meanings, all of which are to various degrees arbitrary and none of which is fully capable of bearing the weight that is placed on it. Nevertheless, insofar as a fundamental task of theological ethics to understand human beings in relation to God, the notion of the image of God is indispensable. In Christian ethics, the image of God is principally understood in two ways. The most common way is to understand it as the ground of human dignity, which in turn provides a ground for certain rights and duties. The image of God thus secures the incommensurable worth of each human being (typically on the assumption that Gen. 1:26f. confers on humans a special status in creation) and the moral requirements that embody respect for that worth. To claim that human nature is a constituent of the image of God in this sense is to claim that certain moral requirements pertain to it by virtue of its participation or inclusion in the image of God. On this account, one task of Christian ethics with respect to the biotechnological determination of human nature is to determine which biotechnological interventions protect or promote the dignity that accrues to human nature, which ones violate that dignity, and which ones are neutral with respect to it. Because human biological nature participates or is included in the image of God as a component of human nature more broadly, this task will be carried out in a way that takes account of the place of biological nature in human nature more broadly, and specifically in a way that recognizes that human nature includes the responsibility of individual human persons to understand the value or purpose of their nature and act accordingly. It will also take account of both the vulnerability of human biological nature and the interdependence of human beings in their vulnerability. Based on these three features of human nature (responsibility, vulnerability, and interdependence), Christian ethics may arrive, respectively, at the standard bioethical principles of autonomy, safety, and fairness.

However, in Christian ethics the image of God can also be understood in a second way that is not inconsistent with the first but differs in emphasis. According to this way of understanding, which has deep roots in patristic theology, to be in the image of God is in some way to conform to Jesus Christ, who is the image of God in the most proper sense (see Rom. 8:29; II Cor. 4:4; Col. 1:15). Conformity to Christ is understood variously as union with Christ, imitation of Christ, and witness to Christ. But whichever of these forms it takes, to claim that human nature is a constituent of the image of God is to claim that it participates in some way in the conformity of the human being to Christ. On this account, the task of Christian ethics with respect to the biotechnological determination of human nature is twofold, namely, to determine the sense(s) in which human nature participates in the conformity of humans to Jesus Christ and to determine whether or in what way(s) biotechnological determination of human nature instantiates or violates that participation. The version of NS4 that I develop in this chapter understands the image of God in this second way, though not to the exclusion of the first way.

According to both these ways, the claim that human nature is a constituent of the image of God means that in some sense it reflects God. God created the human creature to reflect God in the world, and the creature that God created for that purpose is, among other things, a biological creature. It follows that human biological nature is a constituent of the human reflection of God, so that one can, at least in principle, see in human biological nature some reflection of God. The second way holds that this reflection of God is partly constitutive of human life with God: God created humans to reflect something of God; life with God, whether it takes the form of union, imitation, or witness, is lived in that reflection; and their creaturely nature equips humans to live with God in this way. While I consider both these ways of understanding the image of God to be indispensable to Christian ethics, I hold that the second way is the key to the first. As I see it, it is not on the grounds of a prior dignity that humans are eligible for life with God, which consists in union, imitation, or witness; it is rather that the determination of humans by God from eternity for life with God in the form of union, imitation, or witness is the ground of the unique dignity that is theirs as creatures of God. Humans have been created to enjoy a particular form of life with God, and it is this determination of humans as creatures that is the ground of their unique dignity.[2] It is crucial to stress that the determination of humans to image God by union with, imitation of, or witness to Christ is a determination of them as creatures; it therefore pertains to them inalienably, notwithstanding their sinful rejection of it and wherever they may stand with respect to its restoration and ultimate perfection by grace.

By placing the question of the normative status of human nature in the context of the image of God in this second sense, I depart from tendencies of NS1 and NS3 to adjudicate debates over the ethics of biotechnology by direct appeal to a metanarrative about creation and eschatology. As we have seen, NS1 and NS3 differ radically in their understandings of that metanarrative, with NS1 exhibiting an "Augustinian" schema of a finished creation that is eschatologically transformed at the end of time (but not in time) and NS3 exhibiting an "Irenaean" schema of a creation that attains or at least approximates eschatological perfection in time. When these schemas are allowed to directly determine positions on biotechnological enhancement, debates over the ethics of these technologies in Christian ethics are quickly assimilated to an alleged conflict, promoted by some popular writers on the ethics of biotechnological enhancement, between "bioconservatives" who reject biotechnological enhancement in principle and "bioprogressives" who support it in principle. More to the point, however, is that by formulating the normative status of human nature in terms of the image of God instead, I am arguing that the normative significance of human nature concerns first and foremost our life with God, which we live by imaging God. Of course, as such, it also concerns creation and eschatology. But the question I want to pose about the biotechnological determination of human nature cannot be answered by showing how it violates or conforms to the proper ordering of creation and eschatology to one another (which is what NS1 and NS3, respectively, tend to do) but whether it contributes to or detracts from the life with God for which God

---

[2] One advantage of my view is that human dignity so understood does not entail or require the superiority of humans to other creatures; it only entails that humans are determined for a form of life with God that is distinguished by God's having become human in Jesus Christ. Psalm 8 expresses wonder that God has exalted *this* creature, who from a cosmic perspective is so insignificant. Angels are in many ways more impressive creatures than humans, as "posthuman" beings, if they eventually come into existence, might also be, along with beings that might exist, or might have existed, elsewhere in the universe. Yet among these creatures, it is humans that have been created to image God, however impressive or unimpressive they are compared to these other creatures.

has determined us. Of course, our life with God is lived out in the metan-arrative of creation and redemption. But that is precisely because it is life with God who, in Karl Barth's formulation, creates, recon-ciles, and redeems us. What Barth refers to as the history of God's covenant with humanity, which runs from creation through reconciliation to final redemption, is crucial because our life with God unfolds in (and as) this history. But this history is intelligible (and significant) precisely because it is the form that our life with God takes. That life is lived by imaging God as creatures, reconciled sinners, and heirs of redemption. The fundamental question, then, is how human biological nature participates in the image of God, and this chapter attempts to answer this question.

I now turn, in order, to two theologians whose work I consider to be exemplary instances of NS4, namely, Kathryn Tanner and Karl Barth. Neither Tanner nor Barth explicitly formulates the claim that normative status attaches to human nature in the sense that this book understands that claim, and neither engages in debates over biotechnological enhancement. However, both Tanner and Barth can be understood as implicit subscrib-ers to NS4 insofar as human nature (including human biological nature) for them participates in the image of God understood in the second way, namely, as conformity to Christ (union with Christ for Tanner, witness to Christ for Barth). My task is to present their positions in a way that makes explicit their implications for the normative status of human nature in the context of biotechnology.

Tanner on Human Nature and the Image of God

My treatment of Tanner's position here builds on my treatment of it in Chapter 4. We saw there that at the heart of *Christ the Key* is the claim that "God wants to give us the fullness of God's own life," which is not our life until we receive and are remade by the divine Word. The question of human nature concerns the qualities or capacities that are required for this reception and remaking to occur. Tanner articulates these points in terms of human participation in the image of God, and her position can be clarified by looking more closely at the various levels of the image of God that she articulates, drawing on patristic authors. (1) At the highest level is the Word, the second person of the Trinity who shares the divine nature and thus images God as "the perfect manifestation of all that the first per-son is."[3] The Word is thus the image of God in the most proper sense.

---

[3] Kathryn Tanner, *Christ the Key* (Cambridge: Cambridge University Press, 2010), p. 6.

By contrast, humans image God "by participating in what they are not," namely, God.[4] (2) By virtue of its hypostatic union with the Word, however, Christ's humanity "has the divine image for its own" and thus images God perfectly, much as the second person images the first person.[5] Other humans, of course, lack this hypostatic union with the Word. Nevertheless, they image God by approximating it in their own manner as they draw near, through Christ, to the divine image and cling to it. In this sense, they too may be said to image God in a "strong" sense. (3) Moreover, in what Tanner calls a "strong weak sense," humans image God as their creaturely lives are remade in accordance with their attachment to the Word. In these "strong" and "strong weak" senses, humans image God by participation in what they are not, namely, the divine image itself. (4) Finally, humans image God in the plasticity of their creaturely nature, inasmuch as what I have been calling the indeterminacy and open-endedness of their nature resemble God's limitless and incomprehensible nature. For Tanner, this is the image of God in a "weak" sense, as the resemblance to God of human nature as it is in and of itself, apart from participation in what humans are not.

In itself, human nature images God in only this weak sense and thus seems to be of limited normative significance. However, by virtue of its plasticity it plays a key role in the strong weak sense in which humans image God, and this role gives it a much greater normative significance, which I will now elaborate, than it has by virtue of its mere resemblance to God. According to Tanner, it is the almost unlimited capacity of human nature to be shaped by intentional action – that is, its plasticity, which encompasses indeterminacy and open-endedness as well as malleability – that equips humans to receive God's image and be remade by it. The emphasis on plasticity, and the identification of the goodness of created human nature with its capacity to be other than it is, align Tanner with NS3, as we saw in Chapter 4. However, we also saw there how Tanner avoids the most pressing problems with NS3. First, because the plasticity she attributes to human nature is a feature of specific and identifiable human traits and capacities and not just of human nature as such, she can in principle explain why we have the specific created characteristics and capacities we have: They are precisely those that equip us to be remade in the image of God. Although the goodness of our nature consists in its capacity to be other than it now is, it does not acquire new characteristics

---

[4] Ibid., p. 8.
[5] Ibid., p. 13.

or capacities in receiving the divine image and being remade by it. Because the plasticity of the characteristics and capacities with which God has created us enables us to be remade in God's image, our perfection is the perfection of characteristics and capacities given with our creation (however they might have come about in our evolutionary history). Second, the state of perfection has determinately Christian love as its norm, so that the plasticity of our nature is in the service of a substantive norm and not just a vague principle of liberty or progress or the basic or all-purpose goods that equip us for any concrete form of life we might choose.

The crucial point is that Tanner ultimately conceives the plasticity of human nature as its susceptibility to the influence of the Holy Spirit, who works on human capacities so that they become capable of acts or states that go beyond what created nature of itself is capable of doing, even as these capacities, thanks to their plasticity, can remain what they are while being worked over by the Spirit. Tanner's position thus expresses the principle that our nature is equipped for enjoyment of life with God by virtue of the creaturely characteristics and powers that it now has, even as, by virtue of the plasticity of these characteristics and powers, it is susceptible to the work of the Spirit, which takes our nature beyond what is natural to it (that is, beyond what its capacities can do, apart from the work of the Spirit). If normative status attaches to human nature as Tanner understands it (a claim she does not explicitly make) it is not only with respect to its indeterminacy, open-endedness, and malleability (in her terms, its plasticity), as we saw in Chapter 4, but also, and more fundamentally, by virtue of its suitability for our enjoyment of life with God. This is the defining claim of NS4 as I have stated it, and Tanner's position is therefore, at least implicitly, a version of NS4.

At the same time, however, Tanner's version of NS4 is continuous with NS3 in a sense that distinguishes it from Barth's version of NS4. As I noted in Chapter 4, Tanner's theology in *Christ the Key* is deificationist. God wants to give us the fullness of God's own life, but as that is not our creaturely life, we must be taken beyond the state in which God initially created us if we are to enjoy it. Our final state, in which our nature has been worked over by the Holy Spirit, can therefore be described as unnatural with respect to our present state and as divine rather than human. Thus, while normative status attaches to our created characteristics and capacities, which in their plasticity are equipped to accommodate the work of the Holy Spirit, it does not attach to our creaturely nature in its present state, prior to its being worked over by the Spirit. Our created characteristics and capacities remain what they are – by virtue of their plasticity, they already have what they need to be worked over by the Spirit, so that no new characteristics or capacities or supernatural additions to given characteristics or capacities are required – but their plasticity renders them susceptible to a divine action by which we progress from a natural to a divinized state. In other words, by virtue of the plasticity of specific characteristics and capacities of our creaturely nature, we are susceptible to a work of the Holy Spirit that takes us beyond our nature. By contrast, Barth's version of NS4 holds that it is in our creaturely nature in its present state that we enjoy life with God. While this nature is not invariable and unchanging, it does not need to be changed in any way to be rendered suitable for life with God. The Holy Spirit does not work over our nature to render it suitable to life with God but rather works to bring it just as it is into fellowship with God.

What implications does the version of NS4 that I am ascribing to Tanner have for the biotechnological alteration of human nature? Tanner does not address this issue, and no direct implications regarding it can be drawn from her position. However, it is possible to draw some indirect implications that go beyond those I stated in Chapter 3. On the one hand, according to Tanner our natural characteristics and capacities are, by virtue of their plasticity, already susceptible to being worked over by the Spirit. If the Spirit's work can be done without imparting any superadded capacity, it can surely be done without any capacity to be brought about by biotechnological alteration. If so, then Tanner's position offers no positive grounds for claims, of the kind we encountered in Chapter 4, that human perfection is to be pursued through biotechnology. There is no hint in Tanner's position that our present characteristics and capacities are inadequate to the life with God that God wants to give us or that our enjoyment of that life would be facilitated by any alteration of our characteristics and capacities or their replacement by others. On the contrary, Tanner explicitly rejects any such suggestion, asserting that our characteristics and capacities as originally created are already susceptible to the work of grace by which we receive God's life, having been, after all, created for just that end.[6]

On the other hand, it is unclear how Tanner's position can rule out the possibility that the Spirit's work may be accomplished in part through the creaturely agency of biotechnology, as opponents of biotechnological enhancement would wish. To be sure, the Spirit does not have to impart new characteristics or capacities or even to change existing characteristics and capacities (whether by means of superadded qualities or biotechnological alterations), as our nature already has what it needs for the Spirit

---

[6] Ibid., p. 135.

to work it over. However, by working over our existing characteristics and capacities the Spirit does create conditions that are "unnatural" and that could not have been brought about by our characteristics and capacities on their own. To the extent that biotechnology can be understood to generate new, "unnatural" possibilities out of existing human characteristics and capacities (as opposed to generating new characteristics and capacities), it is difficult to know on what grounds Tanner would deny in principle that it could be enlisted in the Spirit's work, as generating unnatural possibilities out of existing natural capacities is just what the Spirit does in working over our nature. In practice, however, it would have to be clear how any biotechnological change promotes the attainment of determinately Christian love that for Tanner is the aim of the Spirit's work. Because Tanner does not say much about what that love consists in, it is not possible to determine which biotechnological interventions, if any, might contribute to it.

We are left, then, with a modest but attractive conclusion, namely, that our understanding of the life with God for which we are created and of the role of our creaturely nature in it offers no principled grounds for supposing that biotechnological determination of our nature might contribute to our attainment or enjoyment of life with God yet does not rule out the possibility that biotechnology might play some role in attaining it. Negatively, Tanner's position rejects the assumption of proponents of biotechnological enhancement that the characteristics and capacities we now have are inadequate to our fulfillment. We already have all that we need to enjoy the life with God that is the purpose for which we were created. The crucial claims regarding creation as a good and finished work are upheld, and the perpetual inadequacy of our creaturely nature that is implicitly or explicitly assumed in arguments for biotechnological enhancement is denied. Positively, the burden of proof is on proponents of biotechnological enhancement to show how attainment of the life with God for which we have been created can be facilitated through biotechnological enhancement. How do the new possibilities generated from our existing human biological functions and traits that biotechnology is capable of bringing about contribute to *this* end? These negative and positive points are of course not made explicitly by Tanner, but they suggest that the fundamental task of Christian ethics in the face of biotechnological enhancement is to determine whether or how biotechnologies contribute or detract from the states and activities that constitute life with God, under the assumption that our creaturely nature is already adequate to that life. That is no small task, as it would require detailed specification of what counts as love of God and neighbor as well as detailed analyses of particular biotechnologies. But it is a task that situates biotechnological enhancement within the theological convictions that define NS4.

Barth on Human Nature and the Image of God

In turning to Barth, we are not so much breaking with Tanner's position as considering a radical alternative to its deificationist orientation. The continuity of these two versions of NS4 is significant. First, for Barth as for Tanner, God wills to give human beings some kind of life with God. For Barth, to be human is to be "determined by God for life with God."[7] Second, for Barth as for Tanner, Christ is the key. He is the one in whom other humans enjoy life with God. Third, for Barth as for Tanner, the question of human nature is the question of the creaturely characteristics and capacities that equip human beings to enjoy life with God. Finally, for Barth as for Tanner, it is as creatures with the characteristics and capacities given us by God with our creation, and not any that are to be acquired or imparted subsequently, that we are determined by God for life with God.

However, these points of continuity are qualified by an important point of discontinuity. In contrast to Tanner, Barth insists that human nature as created by God is already suited to the enjoyment of life with God and need not undergo any working over to be made fit for it. While Barth does not relate this point to technology of any kind, it has important implications for biotechnology (which are detailed subsequently) as it suggests that human nature already has all that is needed to enjoy the form of life with God for which God created humans – not only the characteristics and capacities (as Tanner also holds) but also the natural state in which it was created. The basis of Barth's insistence on this point is his conviction that our determination for life with God was resolved by God from eternity in Jesus Christ as the one who is elected by God from eternity, and has been fulfilled by God in time in Jesus Christ. Because it is from eternity, and thus antecedent to creation, that the human creature is determined for life with God, it stands to reason that the nature of the creature that God brings into existence with the creation of human beings will be suited to this determination. And because the determination has already been fulfilled in time in the incarnate life, death, and resurrection of Jesus Christ, who in his humanity is this creature, it is not still to be brought about by

---

[7] Karl Barth, *Church Dogmatics*, Vol. III, Part 2 (Edinburgh: T&T Clark, 1960), p. 203. This source is hereafter cited as Barth, CD III/2.

any alteration of that nature. It follows for Barth, in contrast to Tanner, that created human characteristics need not be worked over by grace for the determination for life with God to be realized. Rather, what grace enables humans to do is to express these characteristics in actions that confirm what Jesus Christ has already accomplished on their behalf and in their place. It is in that active confirmation that they live their life with God.

Creaturely Nature and the Covenant

We are now prepared to examine Barth's version of NS4 in greater detail. According to Barth, life with God does not consist in sharing the divine life but in a covenant relationship with God. It is this relationship that was resolved by God from eternity and brought about in time. As established and fulfilled in Jesus Christ, the covenant is a covenant of grace. To state it in its most basic terms, in the covenant relationship God is with and for humans in all God's deity, and humans are those whom God is with and for. In Jesus Christ, God takes up the cause of human beings as creatures, as sinners, and as heirs of eternal life, making it God's own cause, and humans are those for whom God so acts. It is in this sense that Barth understands the divine declaration of the covenant relationship: "I will be your God, and you shall be my people."

As resolved by God from eternity in Jesus Christ, prior to creation, the covenant relationship of humans to God has ontological status: The human being is, precisely, the being who is determined by God for this particular form of life with God. "The being [*Sein*] of man ... is the being which is ... drawn into this [covenant] history inaugurated and controlled by God."[8] Human beings do not, for Barth, enter into this form of life with God as those whose being is constituted prior to or apart from it. Rather, as those who are determined in Jesus Christ from eternity for life with God, the covenant of grace implicates their very being. They are ontologically constituted to be God's covenant partners. The question concerning human nature is therefore the question of what *kind* of creature it is who is determined in this way for life with God. What will be the *nature* of this creature whose *being* is constituted by its determination for covenant partnership with God? Because we know that in creating humans, God brought them into existence as beings whose being is this determination for life with God, we also know that God did not create them as beings whose creaturely nature conflicts with their divine determination. On the

---

[8] Ibid., p. 141.

contrary: "If God gives him this determination ... he is obviously ... a being to which this determination is not strange but proper."[9] Elaborating this point, Barth describes human nature as the sign of God's determination of humans for life with God. "Even in his distinction from God, even in his pure humanity, or, as we might say, in his human nature [*Natur*], man cannot be man without being directed to and prepared for [*hingewiesen und vorbereitet*] the fulfillment of his determination, his being in the grace of God, by his correspondence and similarity to this determination for the covenant with God... Consciously or unconsciously, he is the sign [*Zeichen*] here below of what he really is as seen from above, from God. And so he is wholly created with a view to God."[10]

The claim that our creaturely nature is the sign of our divine determination for fellowship with God is integral to Barth's conception of the image of God, and as such it is at the heart of his conception of the normative status of human nature. But before I show how that is the case, it is important to draw attention to two implications of the relationship between our divine determination and the natural sign of it. On the one hand, and in contrast to Tanner, human nature for Barth does not need to become anything other than what it is, as created by God, to enjoy life with God. In its very creatureliness, as created by God, human nature is a sign of humanity's divine determination. To enjoy life with God therefore requires neither supernatural additions to natural human capacities (as Thomists argue), nor an external power of grace that works on natural human capacities (which remain what they are) such that humans are brought to a state that is beyond their nature (as Tanner argues). In this regard, Barth's theological anthropology exhibits in a very strong sense the formula *finitum capax infini*. Life with God is enjoyed in the exercise of our characteristics and capacities as God created them.

On the other hand, this first implication does not tell the whole story. For Barth, God's covenant with us is a covenant of grace. God acts alone, apart from our activity, to establish and maintain us in it. Even apart from sin, human nature has no ability (*Fähigkeit*) or power (*Potenz*) to enter into or remain in covenant fellowship with God. This point (on which Tanner concurs) is captured by a very different formula: *Finitum non capax infini*. The creaturely nature that is ordained to the covenant is one whose characteristics and capacities are suited to active enjoyment of life with God. They are *not* suited to attainment of life with God in the first place

9 Ibid., pp. 205f.
10 Ibid., p. 207.

or to continuation in it. From eternity, God determined that the life of the human creature would be a life constituted by God's grace to the creature, in which God takes up the cause of the creature as God's own cause, and in this second respect, too (that is, in its inability to establish or maintain life with God), human nature is suited to the life with God for which God created it.

Taking these two implications into consideration, we may say that for Barth, just as for Tanner, our creaturely nature, though it requires no supernatural addition and undergoes no transformation to enjoy life with God, is not theologically or morally intelligible in itself, that is, in its creatureliness, apart from its ordination to life with God. In contrast to NS2, it has normative significance not in itself or as such but insofar as it is "the sign here below of what [it] really is as seen from above," that is, from its divine determination for life with God.

Human Nature and the Image of God

With these two implications in mind, let us now look more closely at Barth's claim that our creaturely nature is the sign of our divine determination for life with God. Barth presents our creaturely nature as three-fold, consisting in relationality, as a composite of body and soul, and as temporal. Any one of these three aspects could illustrate the relevance of his account of human nature for technology. His emphasis on the face-to-face encounter as key to human relationality has obvious relevance for the evaluation of a wide range of technologies today, while his insistence on the equal significance of body and soul in constituting human nature are just as relevant to criticisms of certain technologies which turn on the affirmation of embodiment in the face of the reduction of the normatively human to information. However, Barth's treatment of temporality, and more specifically of the human life span, bears on the normative status of human biological nature in a way that is readily apparent and is also of direct significance for a particular form of biotechnology (namely, life-extension technology). For this reason, it is the focus of what follows. My presentation of Barth's position is thus a limited account of Barth's theology of the human creature, but since the point of it is to illustrate NS4 and demonstrate its viability for Christian ethics, and not to expound Barth's theology, it is an acceptable limitation.

Of course, Barth does not discuss life-extension technology in his treatment of the life span, but what he says is relevant to that issue, and I will refer to it as we go along. Barth begins with the apparently unremarkable fact that human life is by nature not infinite but is bounded by birth and death. For him, this fact presents an immediate problem. It seems not to signify our divine determination for life with God but rather to contradict it. The problem he sees is superficially similar to the problems which some proponents of life-extension technologies emphasize, namely, that our desire for life is insatiable, that our aims and projects exceed the time we are allotted to fulfill them, or that life offers us an inexhaustible bounty of goods which cannot be enjoyed in a limited life span. All these supposed problems imply that the good involves a kind of formal infinity with which finite beings can never be equal. They make the dubious assumptions, which (as we saw in Chapter 3) have been questioned by Nussbaum, that infinity is an appropriate ideal for humans and that enjoyment of the good is a quantitative matter. In theological terms, their error is to assume that infinite duration of earthly life is eternal life.[11] By contrast, for Barth the problem arises not out of any ideal of the infinite but out of the specific character of our determination for life with God. It is this determination itself that seems to require endless duration rather than limit. "For man belongs to God. And he belongs to his fellows... What but an unlimited, permanent duration could be adequate for the fulfillment of this determination?"[12] We are determined for covenant relationships with God and our fellow humans, but how, Barth asks, could a bounded life span allow for the fulfillment of *these* relationships? For one thing, God's life is eternal, so how can our bounded lives be adequate to our life with God? Moreover, the fullness of life with God and our fellow humans seems to entail the conformity of our lives to God's perfection, and how could anything short of endless duration be adequate to that? In short, what these concrete relationships offer us and require of us cannot be exhausted in any finite life span, much less in the woefully brief one that is allotted to us as creatures. The very content of the covenant for which we are determined appears to be incompatible with limitation or lack. If that is so, then we must conclude that our bounded life span is not a sign of our determination for life with God but is incompatible with it, so that we are by no means justified in acquiescing in it. Barth, perhaps surprisingly, is sympathetic to this conclusion and endorses, at least provisionally, a principled resistance to our temporal bounds as strongly as any proponent of life-extension technology

---

[11] For a poignant argument against the value of duration of life as such based on both Christian and Stoic premises, see Paul Scherz, "Living Indefinitely and Living Fully: Laudato Si' and the Value of the Present in Christian, Stoic, and Transhumanist Temporalities," forthcoming in *Theological Studies*.

[12] Barth, CD III/2, pp. 555f.

does. "Man cannot abandon the demand that he should endure and burst the limits of temporality."[13] "Resignation [to a limited life span] is incompatible with the fact that life is created by God."[14]

Yet despite the strong case he has just made against limited duration, Barth nevertheless insists that our natural life span *is* a sign of our divine determination for life with God and must finally be accepted as such.[15] His argument rests on Jesus' own humanity as the image of God. Our particular theme is the participation of our biological nature in the image of God and the implications of this participation for the biotechnological alteration of our nature (all of which we are presently considering in the specific context of the biological life span and life-extension technology). What Barth says about this theme is inseparable from a threefold conception of Christ, the image of God, and human nature that coincides roughly with the three levels distinguished by Tanner ("strong," "strong weak," and "weak"), yet he comes to a different conclusion regarding the participation of our nature in the image of God. I will begin by expounding Barth's threefold conception of Christ, the image of God, and human nature; then I will turn in the following subsection to the implications of this conception for the biological life span. But first, I want to state the crucial point of this examination of Barth's position. As I noted at the beginning of this section, Barth holds, in contrast to Tanner, that our nature participates in the image of God just as it is, without having to be worked over by the Holy Spirit in such a way that it is no longer natural with respect to its created state or condition. With this conviction, Barth provides that much stronger a reason to affirm our nature as it is in the face of biotechnological alteration (thus avoiding the problem with NS3) while also relating our nature to something that transcends and embraces it (thus avoiding the problem with NS2).

---

13 Ibid., p. 559.
14 Ibid., p. 555.
15 Barth's insistence on this point is controversial, inasmuch as it entails the claim that death is natural to human beings as created by God and is not simply the result of sin. Barth's position on the relationship of death to our created nature, the implications that follow from it, and the senses in which it is continuous and discontinuous with the positions of Athanasius, Augustine, and Aquinas, have all been discussed with great clarity and insight by Robert Song. See his "Technological Immortalization and Original Mortality: Karl Barth on the Celebration of Finitude," in Philip G. Ziegler, editor, *Eternal God, Eternal Life: Theological Investigations into the Concept of Immortality* (Edinburgh: T&T Clark, 2016), pp. 187–209 (see especially pp. 197–206). As Song emphasizes, Barth's chief concern in naturalizing mortality is to rule out anything in our creaturely nature that could give immortality a foothold there. For Barth, eternal life comes about by resurrection, which is God's act alone. In it, our mortal life is met by God's very different life; it does not involve any capacity for immortality that is in our creaturely nature.

Barth's threefold conception of Christ, the image of God, and human nature is most clearly formulated in relation to two of the three features of human nature he discusses in his theological anthropology in the *Church Dogmatics*. To recall, the three features are relationality (which for him is the most basic feature), body-soul composition, and temporality (to which, as I have noted, the biological life span is fundamental). As I noted previously, my focus is temporality, but because relationality enjoys a kind of primacy over body-soul composition and temporality in Barth's theological anthropology, the participation of temporality in the image of God is best understood in connection with that of relationality, and I will present it in that connection.

Like Tanner, Barth begins his treatment of the image of God with the Trinity. It begins with God as the model or prototype of whom humanity is the copy, made "in the image" and "after the likeness" of God. In what sense is God the prototype to which the human creature corresponds? For Barth, the "Let us make" of Genesis 1:26 implies differentiation within God's own being between the summoning I and the summoned Thou. The human I-Thou relation in turn corresponds to this divine prototype.[16] In keeping with Genesis 1:26f., Barth goes on to identify the encounter of man and woman as the paradigmatic human I-Thou relation and thus the fundamental form in which the human creature corresponds to the divine prototype. As critics of Barth have made clear, this identification is problematic in many respects, and the problems they have pointed out should not be ignored or trivialized. At the same time, it is clear that what is fundamental to Barth's point is a certain form of relationality in which no human being embodies humanity in herself or himself as an isolated individual but only in encounter with the fellow human. Relationality, so understood, can be affirmed without treating sexual difference as the paradigmatic instance of it. Meanwhile, temporality offers a less fundamental but still important aspect of the image of God in which the creature corresponds to God as the prototype. For Barth, God's eternity, in which past, present, and future are simultaneous, is the divine prototype for creaturely temporality, in which past, present, and future occur in inexorable succession.[17] Human beings in their movement from past to present to future are thus the copy that corresponds to God's eternity, in which past, present, and future are simultaneous, as the prototype.

---

[16] Karl Barth, *Church Dogmatics*, Vol. III, Part 1 (Edinburgh: T&T Clark, 1958), p. 196.
[17] Barth, CD III/2, p. 437. For Barth, then, it is nonsuccession of time, not atemporality, that marks God's eternity.

Whether we focus on relationality or temporality, we see here something like Tanner's *weak* sense of the image of God, which has to do with the resemblance of human nature to God. Particular characteristics of human nature – in Barth's case, its relationality and its temporality, and in Tanner's case, its indeterminacy and open-endedness – resemble the divine nature. But for Barth as for Tanner, if we move directly from God to humanity in this way, we get only a weak imaging of God. Like Tanner, however, Barth identifies the image of God in the proper sense with Christ. Yet for him it is not the Word as the second person of the Trinity who is the image of God in the proper sense, as it is for Tanner. Rather, following Colossians 1:15, it is the Word incarnate, "the firstborn of all creation." For Barth, it is in his humanity, and thus in his creaturely nature, that Christ is the image of God in the primary sense.[18] The result is that Barth identifies the two senses Tanner distinguished as *proper* (the Word as second person of the Trinity) and *strong* (the union of Christ's humanity with the Word). For Barth, the Word incarnate does the work of both the proper and strong senses of the image of God in Tanner's schema. I will refer to this composite as the *strong* sense of the image of God. As was the case with the weak sense, I-Thou relationality is fundamental to the image of God in this strong sense as Barth understands it. What characterizes Jesus' humanity for Barth is his absolute and unconditional being for his fellow humans, which accomplishes in time God's absolute and unconditional being for humanity that is God's eternal resolve, and thereby fulfills the covenant of grace. However, God's being for humanity actualized in the humanity of Jesus is itself the repetition of the love that characterizes God's intra-Trinitarian being. "God repeats in this relationship *ad extra* a relationship proper to Himself in His inner divine essence."[19] Thus, "the humanity of Jesus … is the repetition and reflection of God Himself… It is the image of God, the *imago Dei*."[20]

In short, God's eternal intra-Trinitarian love is directed outward to humanity and realized in time in Jesus' love for his fellow humans. In the strong sense, then, the image of God means that in Jesus' humanity, all that God is in Godself, God is also, *ad extra*, for us, and in this way Jesus' humanity perfectly images God. This point is most properly made in terms of relationality, as Jesus' being for us is precisely his relationality. Yet temporality exhibits the same point in a derivative but still important

---

[18] Barth also seems to have Col. 2:9 in mind: "in him the fullness of deity dwells in bodily form."
[19] Barth, CD III/2, p. 218.
[20] Ibid., p. 219.

sense. God's being for us *ad extra* in Jesus Christ is specifically a being in time. God resolves from eternity to give human beings, who are not eternal, a share in God's eternity. To realize this purpose God takes on time, becoming temporal in the bounded life span of Jesus' humanity. And by raising Jesus from the dead, God frees his life in time from its temporal confines and establishes its universal significance; this very time which God has assumed in becoming human is now freed to become God's gift to all human beings of every time. The time of every human being is thus determined by the time God has taken on and given to them.[21] In this way, namely, by virtue of the determination of their own bounded time spans by the bounded time span of Jesus, other human beings participate in their own way in the strong sense of the image of God that, as with Tanner, is fully realized in Jesus' humanity.

Finally, according to Tanner's *strong weak* sense, other humans image God as their lives are worked over and remade in conformity to the Word. They image God as they go beyond their creatureliness and act in unnatural ways. For Barth, by contrast, other humans image God in the strong weak sense as they attest God's being for them in the expression or exercise of their natural characteristics, their creaturely nature (which, we recall, has been brought into existence to equip us for the life with God for which we are determined from eternity) remaining just what it is. Just as for Tanner it is the plasticity of our natural characteristics and capacities that enable our lives to be remade in conformity with Christ, so for Barth it is the way in which our natural characteristics make it possible for Christ to be with and for us that enables us to live in a way that conforms to God's being with and for us in Christ. In both cases our nature is the condition for the life with God for which God creates us. From what we have seen already, it should come as no surprise that for Barth the strong weak sense is found in the relationship of the humanity of other humans to the humanity of Jesus. As an interhuman reality, this strong weak sense remains entirely on the creaturely level as such rather than involving the transformation of creaturely nature by its deification. At issue here is what human nature must be for Christ in his humanity to be for other humans (and for other humans to then live in ways that attest Christ's being for them). How is it that other humans can be represented by Jesus, who in his humanity is for them? In what characteristics of human nature do we find "the presupposition of the fact that Jesus can be for them?"[22] Yet again, relationality

---

[21] Ibid., pp. 439f., 450f., 455f.
[22] Ibid., p. 223.

is fundamental for Barth; it is by virtue of the being of humans *with* one another – their relational nature – that Jesus can be *among* them as the one who is *for* them. Once again, however, human temporality exhibits the same thing as relationality, albeit in a derivative way. Here, it is the limited, bounded nature of human life – the life span that begins with birth and ends with death – that is "the presupposition of the fact that Jesus can be for" other human beings. Boundedness is the characteristic our creaturely nature must have if Jesus in his humanity is to be for us. This brings us back to the matter of our biological life span.

The Biological Life Span as Sign of Our Determination for Life with God

Based on this strong weak sense of the image of God, Barth can now show how our biological life span, which at first seemed to contradict our divine determination, in fact confirms it as "the sign here below" of that determination. As I see it, Barth identifies one general sense in which our biological life span signifies our divine determination and three more specific senses. According to the general sense, if our lives are determined by the gift of God's life in Jesus' humanity, then it is fitting that they should be bounded by a beginning and an end. An unbounded life would still be determined by the gift of God's human life, in which Jesus fulfilled in our place our divine determination for life with God. But it would not attest in its very structure the dependence of our lives on the gracious gift of God's time that determines our lives from outside themselves. As it is, however, the boundedness of our life span refers us to God's grace in precisely this way. In its determination by boundaries – birth and death – that elude our agency, our biological life indicates the dependence of our lives on grace. Bounded as it is by a beginning and end that elude our agency and our control, our life signifies in its very biological structure the dependence of our lives on the gracious gift of Christ's incarnate life that determines our lives from outside themselves, apart from our agency. "By his nature, in virtue of its peculiar character as an allotted span, he is referred and bound to the gracious God as the One who is wholly and utterly outside him but wholly and utterly for him."[23]

---

[23] Ibid., p. 570. As bounded by birth and death, our life span points to God as the limit of our lives. For Barth, this point is crucial to understanding resurrection as the act of God who limits our lives from outside and not as the reinstantiation of an immortality that was original to humans as creatures but was lost with the fall. As Robert Song helpfully points out, while Athanasius, Augustine, and Aquinas held that we were immortal in our prelapsarian state, they denied that immortality was simply a property of our nature. In one sense or another, our prelapsarian immortality depended on

This general observation that the boundaries of birth and death signify the dependence of our lives on God's grace illuminates three more specific senses in which our bounded life span refers us to our divine determination. Two of these senses have to do with our constitution as responsible subjects, that is, as genuine covenant partners with God. First, recall Barth's initial point that life with God seems unintelligible if it involves limit or lack. Life with God and our fellow humans is abundant life, and its very abundance seems to entail endless duration and unlimited time for perfection. Yet, Barth points out, in an unbounded life one would not be confronted with the limits of one's own responsibility for the perfection that life with God demands. Responsibility would be endless, and life would be perpetually *in via* and thus never complete.[24] The fulfillment that seemed to require endless duration would in fact never reach completion. By contrast, a bounded life is one in which we can live in time in the knowledge that the perfection required of our lives has already been accomplished in Christ's time that determines our time. The fulfillment has already occurred, and the very boundedness of our duration signifies our freedom to live our lives as those for whom that burden of accomplishment has been lifted by God. We may generalize this point beyond the case of our own perfection. There is a tendency for humans, at least in the modern West, to assume godlike responsibility for the moral state of the world, taking it upon themselves to conform recalcitrant social and political arrangements to moral norms (as Charles Taylor puts it) and to make history turn out right (as Stanley Hauerwas puts it). This tendency is of course highly commendable in many regards, and few morally serious people would welcome its demise. But it is also ambiguous in ways that Taylor, Hauerwas, and many others have pointed out and as victims of moral progress attest. The principle of limited responsibility that applies to our own perfection also applies to perfection on this larger scale; in both cases, the point is that our responsibility is to be formed by the awareness that God has taken responsibility for us. This awareness must be combined with another form of awareness indicated in the second of Barth's three points (which we will consider shortly) to avoid forfeiting the gains of

grace. Barth differs from them only in denying immortality any foothold in our creaturely nature. For Barth, resurrection, not immortality, is our hope, and resurrection is the act of the God who limits our bounded lives from outside, not an act that involves something in our nature, such as the soul, that by God's grace makes us immortal. See Song, "Technological Immortalization and Original Mortality," pp. 204f.

[24] Barth, CD III/2, pp. 561f.

moral activism in favor of moral quietism, but it plays an important role in avoiding the characteristic dangers of moral activism.

This point has an important implication for the ethical evaluation of life-extension technology. We saw in Chapter 3 that critics of the quest for unending life such as Nussbaum and Kass argue that an indefinitely long life would be frivolous. Without the need to face impending death we would lack the resolution to accomplish anything significant for ourselves or others. But proponents of radical life extension often appeal to a strong moral motivation for living longer. They stress how much more a human being could accomplish if he or she could live for centuries rather than for mere decades. They envision lives of godlike accomplishment, which presupposes godlike responsibility for our lives and the life of humanity in general. By contrast, a bounded life is one in which our biological nature reminds us that it is not up to us to accomplish all that can be and must be accomplished for our moral vocation to be fulfilled and our perfection accomplished. A bounded life turns out to be a condition for living by God's grace.

The second sense in which our bounded life span refers to our divine determination involves another feature of moral responsibility that complements the one we have just been considering. An unbounded life would lack the specificity without which one cannot be a subject and thus a genuine covenant partner. In one's birth and death, Barth points out, one is "irreplaceable, indispensable, and non-interchangeable." In these two events, one cannot be represented or replaced by another; unlike all other events in one's life, here one is uniquely, unqualifiedly, and unalterably oneself.[25] These two events thus give the life that runs from the one to the other the singular identity that is a necessary condition not only of the incommensurable dignity of the person, who cannot be exchanged for any other person, but also of responsibility, which requires one to answer for oneself. They also endow the particular, the contingent, and the temporary, which would be meaningless in an unbounded life, with worth. These factors invest responsibility with the urgency that is the most commendable feature of modern moral activism. Finally, the singular identity that is constituted by one's irreplaceability in one's birth and death is also the basis for the uniqueness of one's vocation, which is that aspect of the moral life that is not resolvable into the universality of moral requirements. A life span that is bounded by birth and death gives the life that runs from

[25] Karl Barth, *Church Dogmatics*, Vol. III, Part 4 (Edinburgh: T&T Clark, 1960), pp. 569–85. This source is hereafter cited as Barth, CD III/4.

the one to the other the singular identity that is a condition for a specific human vocation, with the limited yet intensive ethical responsibility it involves.[26] Once again, this point has an implication for life-extension technology. Some proponents of radical life extension are enticed at the prospect of nearly limitless possibilities for self-transformation. In the bounded life that we live, however, to think and act in this way keeps us from seizing the particular vocation that is available to us.

Finally, the third sense in which our biological life span signifies our divine determination for life with God is an indirect one (which for Barth derives from the second sense) but is still important.[27] It involves the life span as a life cycle in which, as a matter of course (though of course not in the case of every individual human being), birth is followed by growth and death is preceded by decline while between them lies a middle period of greater or lesser length. Proponents of radical life extension do not typically argue for a lengthening of the first or last stages of the cycle (indeed, the risk that living longer would involve a long, drawn-out decline is acknowledged as a serious one that life-extension technologies would have to surmount). Their favored scenarios involve a vastly extended middle stage of life in which physical strength, mental acuity, and emotional vitality remain optimal for great lengths of time while the duration of decline is ideally reduced but at least not increased and the stage of growth remains what it now is. Whatever we might think of the biological viability of these scenarios, a greatly extended middle stage of life – lasting, let us say, for a century or longer – would be one in which one could largely if not almost entirely remove the stages of growth and decline from one's view. This would amount to a life in which the characteristic attitudes, practices, and attachments to others that mark the middle stage of the life span would be formed and exercised without awareness of the other two stages. As it now is, of course, the limits of the life span readily facilitate the awareness of the stages of growth and decline that forms these attitudes, practices, and attachments. We can, if we choose, quite easily remember the stage of

---

[26] Ibid., pp. 595–607.

[27] See ibid., pp. 607–18. My discussion of this third sense is indebted to Gilbert Meilaender's analysis of radical life-extension, and in particular his contrast between duration (the length of one's life) and shape (its unfolding in stages), and to Autumn Ridenour's presentation of Barth's treatment of the stages of human life. See Gilbert Meilaender, *Should We Live Forever? The Ethical Ambiguities of Aging* (Grand Rapids, MI: Eerdmans, 2013); Autumn Ridenour, "The Coming of Age: Curse or Calling? Toward a Christological Interpretation of Aging as Call in the Theology of Karl Barth and W. H. Vanstone," *Journal of the Society of Christian Ethics* 33 (2013): 151–67; and idem, "Union with Christ for the Aging: A Consideration of Aging and Death in the Theology of St. Augustine and Karl Barth," unpublished PhD dissertation, Boston College (2013).

growth and anticipate the stage of decline; when we do, the result is that the attitudes, practices, and attachments that characterize the middle stage are in part shaped and qualified by the other stages. To take a straightforward example (albeit a somewhat bourgeois one), one's satisfaction in one's middle-stage achievements is qualified by gratitude for all of those who during the stage of growth made one who one is and by humility in the anticipation of the stage of decline in which one will have to depend on others to pass on or continue what one has accomplished. One might plausibly argue that these very specific forms of expression of gratitude and humility are essential to living the middle stage in a worthy manner, yet if the middle stage were to last, say, a century or two, our biological nature might make it difficult to cultivate these attitudes, practices, and attachments and might favor the cultivation of very different ones. The point, however, is that a biological life span that unfolds in stages equips us for a life in which the characteristics of each stage are formed in awareness of the other stages.

Before turning more concretely to its implications for the ethics of biotechnological enhancement, I will briefly summarize Barth's version of NS4. In Jesus Christ, human beings are determined by God from eternity to be those whom God is with and for, and this determination is fulfilled by Jesus Christ in time. As such, Jesus Christ is the perfect and proper image of God: He is the one in whom God is God with and for humanity. Other humans image God as those whom God is with and for in Jesus Christ: In their life with God they attest that they are those whom God is with and for in Jesus Christ. As God has determined humans for this particular form of life with God, God creates them with a nature that equips them to live as those whom God is with and for. Their very nature is thus a sign of their determination by God; as such, it images the God who is with and for them in Jesus Christ. The biological life span, with its boundedness and stages, is one aspect of their nature that signifies the determination of humans for this form of life with God and equips them for it. I have identified one general and three specific ways in which humans, equipped by their biological life span, are enabled to live in a way that images God by attesting that they are those whom God is with and for.[28]

---

[28] The Barthian version of NS4 I have just presented is admittedly incomplete. I have merely explained how humans are equipped by one aspect (namely, temporality) of their threefold creaturely nature to conform to Christ by living as witnesses to God's being with and for them in him. Even with respect to this one aspect of human nature, I have not examined what the life of witness consists in, which would require us to go beyond creation to consider how it is inflected by the reconciliation of sinful human beings in Jesus Christ and their anticipation of final redemption (and with it, eternal life)

Implications of Barth's Position for Biotechnology

The version of NS4 I am drawing from Barth attaches normative status to human nature as it signifies in its characteristics and capacities God's being with and for humans in Christ and equips humans to live as those whom God is with and for, thereby conforming to Christ. Life with God, so understood, is therefore lived in their nature as God created it, and not in a nature that is still to be brought about. By focusing on the biological life span, I have demonstrated the viability in principle of this version of NS4 as a stance of Christian ethics toward the biotechnological determination of human nature and have identified its general implications for life-extension technology, which is an important and (for many people) appealing line of human biotechnological research. However, I have not yet clarified how this version of NS4 would evaluate the biotechnological determination of human nature, which in accordance with our focus would mean life-extension technology. This clarification will require extrapolation from Barth's actual position, which of course did not take the prospect of radical life-extension or any other biotechnological alteration of human nature into account.

The first thing to be said is that Barth's position offers even stronger grounds than does Tanner's for the claim that human nature already has what is needed for humans to enjoy life with God. For Barth, not only is it the case that our characteristics and capacities are sufficient as they now are (as Tanner also holds); it is also the case for him that life with God is not enjoyed in a state, to be brought by the work of grace acting on our characteristics and capacities, that is unnatural with respect to our nature as initially created by God, but rather in the exercise or expression

in him. In particular, nothing has been said about how one's limited yet intense responsibility is to be directed, what one's singular vocation should be, or what attitudes, practices, and attachments constitute a worthy life at each stage. Furthermore, this picture is not only incomplete; it is also in one key respect a distorted one. For Barth, our actual lives are lived as reconciled sinners, not as creatures pure and simple, so that even a more complete account of our creaturely nature and its inflection by reconciliation and redemption would be inadequate without presenting this entire account from the perspective of reconciliation. Nevertheless, while the picture I have sketched of a Barthian version of NS4 is radically incomplete and partly distorted, it is not inconsequential. For Barth, the life we live as reconciled sinners is lived as the creatures we are. More precisely, our creaturely nature is the "presupposition" of our life as reconciled sinners and heirs of redemption; it is the condition in which our life with God in those domains is lived. What happens with and to our creaturely nature is therefore of the greatest significance, and it is our creaturely nature that is directly implicated by biotechnology. The effects of biotechnology on our relationality, body-soul composition, and temporality have to do with the conditions in which we live with God and one another as reconciled sinners and heirs of redemption. In light of these factors, the risks of incompleteness and distortion are well worth taking.

of our natural characteristics and capacities – that is, in the actualization of possibilities that are inherent in these characteristics and capacities as they have been created by God. Grace activates, empowers, and directs the actualization of these inherent possibilities, enabling us to perform actions that attest God's being with and for us that would otherwise be impossible for us to perform. To say it once again, the capability is from grace, but the capacities that grace activates and directs are natural. For Barth, then, life with God does not take us beyond our created state. His position thus precludes the potential justification of biotechnological alteration of human nature that Tanner's position leaves open. It offers no ground for claims that biotechnology might be employed in a work of grace that generates possibilities out of human characteristics and capacities that are not already inherent in human nature as created by God.

At the same time, it is unclear that this position rules out biotechnological alteration of human nature altogether. The notion of possibilities that are inherent in human nature as created by God, and of the relationship of biotechnology to those possibilities, can be conceived in two ways, both of which preserve the crucial points that human nature already has the characteristics and capacities that equip it for life with God and that life with God is lived in our nature as God created it. It can mean, first, that all the possibilities that become actualized in the enjoyment of life with God are given in an initial state of human nature as created by God. In this case, human nature as created by God is conceived as fixed or determinate in the manner of less sophisticated versions of NS1 and NS2, and whatever new possibilities biotechnology might introduce would contribute nothing to the enjoyment of life with God and may detract from it. But it can also mean that human nature as created by God is, as NS3 holds (and as sophisticated versions of NS1 and NS2 allow), indeterminate, open-ended, and malleable in such a way that its inherent possibilities are not only those of its initial state. In this case, while human characteristics and capacities remain what they are (the crucial point, which distinguishes NS4, with its convictions regarding creation as a finished work, from NS3), a full account of the possibilities that are inherent in them could include at least some that biotechnology may draw out of them, so that, in their case at least, biotechnological alteration does not bring about an unnatural state with respect to human nature as created by God but rather expands or reconfigures possibilities that are inherent in human nature as created by God.

While Barth's account of human nature is not incompatible with either of these alternatives, it readily accommodates the second one. The features of human nature to which in his account normative status attaches (namely, relationality, body-soul composition, and temporality) are general, and their normative significance (that is, their role as signifying our life with God and equipping us for it) is compatible with a wide range of concrete variation in the manifestation of these features. This point is clear in the case of the biological life span, which has been our focus. Our temporality would signify our determination for life with God and equip us for it at any length of life span, however long or short, so long as that life span remains bounded by birth and death and takes shape in stages. Its normative significance, then, is compatible with the range of possibilities envisioned in all but the most extreme life-extension scenarios.[29] Most proposals for radical life-extension anticipate a life span that is exceedingly long, yet still bounded. Aside from its commitment to the life cycle with its three stages, the version of NS4 that I draw from Barth seems to have no direct objection to such proposals, unless it is a warning not to cross the line into actual immortality (assuming that could ever be done) and thus eliminate one boundary of our lives. Of course, the commitment to the life cycle requires the preservation of the three stages, and to that extent it is in tension with scenarios that envision a greatly extended middle stage followed by a much-abbreviated decline and death.[30] But it does not altogether rule out such scenarios, so long as they meaningfully preserve the three stages even as they significantly extend the middle stage.

In short, it would be exceedingly difficult on Barth's account for biotechnological alteration of our nature to efface the sign our nature offers of our determination for life with God or render it unsuited to equip us for life

---

[29] Although mainstream advocates of radical life extension sometimes use the language of immortality, they typically refer to the expectation or hope that science and technology will progressively eliminate the causes of cell and tissue degeneration to the point that humans live dramatically longer lives and death is indefinitely, though not infinitely, forestalled. They do not argue that eventually humans will no longer die. See Aubrey de Grey, *Strategies for Engineered Negligible Senescence: Why Genuine Control of Aging May Be Foreseeable* (New York: New York Academy of Sciences, 2004); Aubrey de Grey and Michael Rae, *Ending Aging: The Rejuvenation Breakthroughs That Could Reverse Human Aging in Our Lifetime* (New York: St. Martin's Griffin Press, 2007); Laurence D. Mueller, Cassandra L. Rauser, and Michael R. Rose, editors, *Does Aging Stop?* (New York: Oxford University Press, 2011). Of course, virtual immortality has been proposed by popular advocates of radical life extension who place their hopes in the eventual merger of human intelligence with computers. These proposals are what I mean by speaking of the most extreme life-extension scenarios. See Ray Kurzweil, *The Age of Spiritual Machines: When Computers Exceed Human Intelligence* (New York: Penguin, 2000), pp. 101–31; and Hans Moravec, *Mind Children: The Future of Robot and Human Intelligence* (Cambridge, MA: Harvard University Press, 1988), pp. 100–24.

[30] These so-called compressed morbidity scenarios, in which decline is abbreviated almost to the vanishing point, are generally thought to be unrealistic because it is unclear what one would die from if one were healthy and vigorous until just before one dies.

with God. Specifically, if the biological life span possesses the normative significance that this version of NS4 attributes to it, then even quite dramatic life-extension technologies would not imperil that significance. So long as human life is bounded by birth and death and unfolds in roughly the three stages, it appears to play its normative role, however long or short its span. The obvious conclusion, at least in the case of temporality but presumably also in the cases of relationality and body-soul composition, is that Barth's position could accommodate in its account of the possibilities of human nature as created by God many of the possibilities that biotechnology might bring about without imperiling its claims about human nature as signifying and equipping us for life with God and as sufficient for life with God in its present characteristics and capacities. To state this conclusion positively, while we have the characteristics and capacities we need to enjoy life with God (that is, relationality, body-soul composition, and temporality), the possibilities inherent in those characteristics and capacities need not all be given in an initial creation of humans by God and need not be fixed but could be expanded and reconfigured by various factors, including biotechnology. If this is so, then there is no determinate set of possibilities that we need simply accept by default as bearers of the normative significance of human nature. Rather, we are free to ask which possibilities most properly instantiate that normative significance, and the answer to this question presumably might include possibilities that are brought about by biotechnology.

However, this conclusion is far from decisive. Recall that human nature signifies and equips us for our life with God who is with and for us in Jesus Christ and that we live that life as we attest God's being with and for us in Jesus Christ in our actions, thereby imaging God, whose being with and for us in Jesus Christ is reflected in our actions. In the case before us, we image God in this way by acting as those whose lives are bounded, whose responsibility for the perfection of our own lives and the world around us is intense but limited, whose vocations are singular and determinate, and who live each stage of our lives in vivid awareness of the other stages. Barth makes clear how our actions are to reflect this purpose of our nature in their actualization of already-given possibilities of our nature (in this case, the life span as it now is). The question before us, which Barth did not consider but that biotechnology poses for us, is how our actions are to reflect this purpose of our nature in the context of the generation of new possibilities of our nature (in our case, new possibilities of duration and shape of the life span). This question breaks down into two questions, one concerning the effect of our actions on our nature and the other concerning our actions themselves as attestations of the meaning and purpose of our nature. So, first, do the alterations of our natural characteristics and capacities instantiate the meaning and purpose of those characteristics and capacities? In our example, does a greater duration or different shape of life span better instantiate God's being with and for us in Christ's life lived for us? And, second, do the actions by which we alter our characteristics or capacities attest our life with God? In our example, in acting to extend the life span or alter its shape are we expressing the meaning purpose of the bounded life span and thereby attesting God's being with and for us in Christ's life lived for us or are we subverting it? Both questions are important. If God as Creator does not fix in advance the possibilities of the characteristics and capacities that bear the normative significance of human nature but accommodates creaturely action (including, but not only, our intentional action) in their determination, then it is necessary to judge which if any of the possibilities of biotechnological alteration of human nature instantiate that normative significance. And if our action is to image God, reflecting God's being with and for us in Christ, then it is necessary to judge which if any of our acts of biotechnological alteration of human nature witness to Christ in that way.

In the case before us, it is difficult to see how an affirmative answer can be given to these two questions. As for the first question, the point of the biological life span, as we have seen, is inseparable from its boundedness and three-stage shape, and there is no reason to suppose that the new possibilities that biotechnology might be capable of – possibilities that involve extending the length of the life span or of its middle stage relative to the other stages – would instantiate the point of the life span better than its current possibilities (that is, its current length and shape) do. Indeed, to the extent that these alterations defer death and decline, they may instead instantiate a diminished significance of boundedness and shape that undermines the meaning and purpose of the life span. Turning to the second question, it is unclear how the human acts by which the life span or its middle stage is extended would attest the imaging of Christ that is the point of the biological life span and thus also of the actions that determine its duration and shape. In acting to extend life, how is one expressing in one's action one's awareness of the dependence of our lives on Christ's life lived for us, cultivating limited yet intense responsibility, and forming attitudes and dispositions appropriate to the stages of the life span? It appears that in so acting, one is expressing in one's action the meaning and purpose that attach to duration of life rather than those that attach to its boundedness and shape and thus attesting something other than Christ's being with and for us in his bounded life span. In sum, neither the altered life span nor the action of altering it appear to image God in the strong weak sense.

Thus would the version of NS4 I have drawn from Barth evaluate one prospective biotechnological alteration of human nature. However, the negative judgment rendered in this case should not be generalized to other cases or even taken as the final word in this case. Generalization is unwarranted because, in the first place, a focus on another characteristic of human nature to which normative status attaches, such as relationality, might yield different conclusions regarding the justifiability of biotechnological alteration. It is not inconceivable that some enhanced state of a certain aspect of our social nature might, at least in principle, instantiate the normative significance of relationality and that the action that brings it about might at the same time attest that significance. In the second place, as for the finality of the negative judgment on radical life-extension, a different verdict might emerge if temporality is not evaluated alone, as if it were independent of the other characteristics to which normative status attaches, but in relation to those characteristics. For example, respect for human nature as a body-soul composite might justify a program designed to eliminate conditions that threaten bodily life, with the unintended but foreseen (and perhaps welcome) result that the duration of the life span or its middle stage is significantly increased. Participation in such a program would not necessarily instantiate or attest the problematic meanings and purposes, mentioned previously, that programs aimed at life-extension inculcate.

To conclude, this version of NS4 evaluates biotechnological alteration of human nature by asking whether it actualizes the meaning and purpose of human biological nature and thereby images God, and it raises this question with regard to both the effects of biotechnological action on natural characteristics and capacities (do the altered characteristics and capacities better instantiate their meaning and purpose?) and to the actions by which the alterations are made (in so acting, am I inculcating the meaning and purpose of the relevant characteristics or capacities and thus attesting it in my action?). The crucial claim is that the point of our creaturely nature is to equip us to image God by attesting God's being with and for us in actions as creatures. The question posed to biotechnology, then, is this: In acting to determine the possibilities of our creaturely nature, are we instantiating (in our creaturely nature as well as in our actions toward it) that point or some other one?

Conclusion

As NS4 is my construction, and will be unfamiliar even to readers who follow biotechnology debates closely, it is appropriate to conclude this chapter with a summary of it along with an evaluation.

The defining claim of NS4 is that normative status attaches to human nature as the condition for a particular form of life with God for which God created human beings. God created humans with the natural characteristics and capacities they have because it is these characteristics and capacities that suit them for the life with God that was God's purpose in creating them. For Tanner, life with God is realized *through* our natural characteristics and capacities, which are worked over by the Spirit in order that we may become what we are not, namely, sharers in God's own life. For Barth, life with God is realized *in* our natural characteristics and capacities, which equip us to live as those whom God is with and for and which are expressed as the Spirit activates, directs, and empowers actions that confirm God's being with and for us. But in either case, NS4 approaches the prospect of biotechnological alteration of human nature with the premise that our nature already has what is needed for humans to enjoy life with God. We are not waiting for cognitive, perceptual, emotional, or physical enhancements that will make it possible for us to enjoy a kind or manner of life with God that our present characteristics and capacities render us unable to enjoy. To enjoy life with God it is not necessary for us to acquire different characteristics or capacities than those we now have (such as new physical or emotional characteristics or new cognitive or perceptual capacities) or even to augment the characteristics and capacities we now have. Contrary to NS3, then, for NS4 the perfection of our nature is not attained along the line of its biotechnological enhancement. Rather, it is realized in the life with God that we enjoy as grace acts on the characteristics and capacities that have been given to us with our creation by God. Our created characteristics and capacities are therefore adequate to the life with God for which God has created us, and this adequacy is implied by the claim, as NS4 understands it, that creation is a good and finished work. By whatever processes our characteristics and capacities may in fact have come into existence, they are the characteristics and capacities that suit us for the life with God for which God created us, and we enjoy that life as grace acts on them to bring us into it. Finally, the ground for the claim that we enjoy life with God in or through our present characteristics and capacities is Christological. Because Jesus Christ both realizes life with God in its fullness in his humanity and is the condition by which we enjoy life with God in or through our humanity, we need no other characteristics or capacities than those that constituted his humanity to enjoy life with God.

At the same time, NS4 denies that the normative significance of human nature is grounded in human nature. It does not consist in the fulfillment of aspects of our creaturely nature simply as such, as subscribers to NS2 hold, but rather in the role of our creaturely nature in relating us to something that both transcends and embraces it, whether that is conceived as God's own life (Tanner) or God's being with and for us (Barth). In either case, human nature as created by God suits humans for a particular form of life with God, and it is only when we understand what that life with God is that we can understand the normative significance of human nature and can identify the feature(s) of human nature with respect to which normative status attaches to it. For Tanner, life with God is sharing in God's own life, and that means participation in what we are not. It requires us to become something that exceeds the inherent possibilities of our nature as created by God, and the normative significance of our nature as created by God lies in the susceptibility of our capacities to the work of grace that takes us beyond what is natural to us as creatures and thus makes us capable of participation in what we are not. For Barth, life with God is fellowship with the One who is with and for us in all the divine being, who establishes and maintains us in that fellowship, and who has created us with a nature that is suited to it. The normative significance of our nature is found in the ways it signifies God's being with and for us and equips us to live as those whom God is with and for, while we do live as those whom God is with and for as grace activates, empowers, and directs creaturely actions that attest God's being with and for us.

Neither the version of NS4 that I have drawn from Tanner nor the one that I have drawn from Barth rules out the possibility that biotechnological alteration might contribute to life with God as Tanner and Barth understand it. In Tanner's case, that possibility would imply that biotechnology may be a means by which humans are brought beyond their created state by the work of grace on their existing created characteristics and capacities which, by virtue of their plasticity, are susceptible to such a work. To the extent that biotechnology generates new possibilities out of existing characteristics and capacities, and to the extent that these possibilities instantiate or contribute to the love of God and others in which humans image God, Tanner's position does not appear to rule out biotechnological alteration of human functions and traits, though it remains unclear what it would mean for biotechnology to meet the second of these two conditions and whether any particular biotechnologies would in fact meet it.

In Barth's case, the normative significance of the human characteristics and capacities with respect to which normative status attaches to human nature (namely, relationality, body-soul composition, and temporality) can accommodate considerable variation in those characteristics and capacities. It is not the case, then, that the nature that equips us for life with God is a fixed nature, such that any biotechnologically induced change to these characteristics would render them unsuited to the realization of life with God. Rather, the pressing question is whether any such biotechnological alteration would instantiate the point of creaturely nature (that is, its meaning or purpose as created by God). For Barth, we live our life with God as we image God by attesting God's being with and for us in our creaturely action. In acting to determine the possibilities of our nature, are we actualizing this purpose of our nature or some other purpose that might subvert this one? To return yet again to the case at hand, in acting to extend the life span or its middle stage, does our extended life better instantiate the boundedness of human life by grace and the limited yet intense responsibility, singularity and determinacy of vocation, and awareness of other stages it involves? And does our action inculcate and express these meanings, thereby attesting God's being with and for us in Christ's life span? Or is some alternative meaning and purpose instantiated and attested, so that our nature and our action image something other than God's being with and for us in Jesus Christ?

In considering the possibility, which both Tanner and Barth leave open, that biotechnological enhancement might contribute to the role of our nature in equipping us for life with God, it is important to keep in mind that for both Tanner and Barth, fulfillment of this role is not indexed to greater and lesser qualitative or quantitative states of our characteristics and capacities. To equip us for life with God, it is sufficient that we have the characteristics and capacities that Tanner and Barth identify; their role in rendering our nature suitable for life with God does not depend on meeting qualitative or quantitative thresholds. For Tanner, it is sufficient that we possess rational, volitional, and sensuous capacities with the plasticity that characterizes them in their distinctively human form; for Barth, it is sufficient that we have characteristics such as sociality, body-soul composition, and temporality. In neither case is the suitability of our nature for life with God indexed to, say, degrees of cognitive ability or a certain duration of life span. It follows that whatever case might be made for the contribution of biotechnological enhancement to the meaning and purpose of our biological nature will not be made on the grounds that the relevant intervention renders our characteristics or capacities better suited to life with God than they now are.[31] If in Tanner's case the Holy Spirit might make use of such interventions to draw possibilities out of our existing capacities that promote love of God and others, and if in Barth's case biotechnological changes to our life span might express its point more adequately, what is at stake is the actualization of possibilities of a nature that is already equipped for life with God, and not anything that might render our nature better equipped.

As these considerations make clear, the conviction that we already have the characteristics and capacities we need to enjoy life with God looms large over NS4. The case for biotechnological enhancement rests strongly on the assumption that our natural characteristics and capacities are inadequate to the meaning or purpose of our lives as creatures; by insisting on the opposite, Tanner and Barth force proponents of biotechnological enhancement to justify the latter in terms of its contribution to a meaning or purpose for which our nature is already adequate and that is established prior to and apart from intentional biotechnological intervention. If the point of our natural characteristics and capacities is to equip us for life with God, the questions for biotechnology are whether the alteration(s) in question contribute to or detract from the life with God for which our nature equips us and whether the action(s) of altering them instantiate the point or subvert it. If, for example, I act to extend my life span, does the extension instantiate the threefold meaning of the bounded life span, and does my action inculcate these meanings or betray them?

If these features of NS4 render it a viable option for Christian ethics as it encounters biotechnological enhancement, it does not follow that Tanner's and Barth's particular versions of NS4 are without flaw. As I will now show, neither version avoids entirely the problems that cling to the alternatives to NS4 that they most closely resemble. For Tanner, the normative significance of our creaturely nature consists entirely in its capacity to be transformed into something else. There is no value in our present state as such, and the value of our present characteristics and capacities lies in their

---

[31] Although I do not develop it here, this point has important implications for the issue of disability. If the realization of the meaning and purpose of human nature requires only the incidence of distinctively human characteristics and capacities, and not a qualitative or quantitative threshold of expression or functioning of these characteristics and capacities, then even people with severe cognitive, emotional, or physical impairments have the nature that is needed for life with God. From this perspective, to look for the perfection of human nature in the qualitative or quantitative changes to human characteristics and capacities that biotechnological enhancement may be capable of delivering is to reflect and perpetuate ableism.

capacity for the transformation of our nature into another state. Such a transformation is necessary if we are to share in God's own life; to do that, we must become what we are not. But these considerations suggest that Tanner's position is ultimately a version of NS3 and is subject to some of the limitations of NS3, chief among which is its insufficient affirmation of the goodness of our nature in its present state. It would be difficult to derive from Tanner's position an ethic of concern for the well-being of our nature as such, with the possibilities that are inherent in it apart from the work of the Holy Spirit. This is not to say that an adequate ground for securing the well-being of nature in its present condition cannot be found elsewhere in Tanner's theology. But the absence of such a ground in her version of NS4 is clearly a shortcoming.

For his part, Barth risks the opposite error. For him the point of human nature appears to be fully realized in its present state. Life with God is enjoyed entirely in the possibilities that are inherent in our nature as God created it. This position accords great value to our creaturely nature. It grounds a moral stance that (as any careful reader of Barth's ethics of creation will discover) enjoins respect and care for human beings in their vulnerabilities and limitations and is inimical to all norms and practices that denigrate or spiritualize human nature or are insufficiently solicitous of its well-being. Barth's doctrine of the human creature has been rightly described as a celebration of the goodness of human finitude.[32] However, his position incurs a major theological liability. While it may be a defensible position regarding creaturely nature prior to the eschaton (according to I John 3:2, "what we will be has not yet been revealed"), it is questionable whether its insistence that our life with God is lived entirely in (and not just through) our nature as created by God allows for any eschatological transformation of our nature into the divine likeness (according to the same verse, "when he is revealed, we will be like him"). To the extent that Barth considers our resurrected nature, he tends to depict it as the same as our nature now: In the resurrection, our finite life remains as it is but is somehow met by the very different life of God.[33] It is unclear in what sense this condition would count as eternal life, which would seem to require some transformation of our created nature, and not merely its encounter with God, and it comes perilously close to the notion of resurrection as a kind of resuscitation. These considerations suggest that Barth's position is

---

[32] See Fergus Kerr, *Immortal Longings: Versions of Transcending Humanity* (London: SPCK, 1997), pp. 23f. Quoted in Song, "Technological Immortalization and Original Mortality," p. 207n5.)

[33] See, e.g., Karl Barth, *Church Dogmatics*, Vol. IV, Part 3.1 (Edinburgh: T&T Clark, 1961), pp. 310f.

not as significant a departure from NS2 as its conviction that the meaning and purpose of human biological nature are grounded in something that transcends us implies. Like Kass and Nussbaum, Barth finds the meaning and purpose of human nature in vulnerabilities and limitations of biological nature (such as the boundedness and shape of the biological life span). Even though for him that meaning and purpose do not inhere in a purely immanent struggle with vulnerability and limitation, but rather in the reference of the relevant characteristics to God's being with and for us, there is no place in his position for the notion that the eschatological destiny of our biological nature exceeds or changes the possibilities that are now inherent in it or that such a prospect has any implications for what it means to image God in our creaturely lives this side of the eschaton.

Another limitation of Barth's version of NS4 has to do with the three characteristics of human nature (relationality, body-soul composition, and temporality) he identifies. While these characteristics are neither arbitrary (Barth carefully derives them from his conception of our nature as the creaturely condition of our determination for life with God) nor exhaustive (Barth leaves room for "phenomena" of human nature that, once our determination for life with God is established, can and must be affirmed as genuine signs of that determination), they are in an important sense abstract.[34] They have to do with human nature as related (relationality), integrally whole (body-soul composition), and situated (temporality) but not with the concrete functions or the rational, volitional, and sensible capacities that are the "stuff" of human nature and in which Tanner finds the plasticity that equips human nature for life with God. As a result, it is unclear how these concrete aspects of our created nature signify its meaning and purpose, and therefore unclear as well what implications biotechnological intervention into them would have for that meaning and purpose. I have demonstrated the relevance of Barth's conception of temporality to biotechnological extension of the life span and have suggested that his conceptions of relationality and body-soul composition might be relevant to technologies involving our sociality and embodiment. But the overall relevance of Barth's version of NS4 to the ethical evaluation of biotechnological enhancement will depend on the extent to which his position allows for concrete functions and capacities to be explained as constituents of relationality, body-soul composition, and temporality.

The flaws of these two versions of NS4 should not be dismissed or trivialized. Nevertheless, that NS4 can accommodate positions with opposite

---

[34] On the "phenomena" of the human, see Barth, CD III/2, pp. 71–132.

shortcomings such as these is not entirely to its discredit; it indicates that it is broad enough to include the range of plausible theological positions that fall between Tanner's and Barth's positions. In any case, the vices of these two positions do not nullify the virtues of NS4, which by considering our nature in relation to its purpose in our life with God is able to avoid the weaknesses of its rivals while also accommodating their strengths. This last point provides occasion for considering a final shortcoming of NS4, namely, its restriction to Christian ethics. Subscribers to NS1, NS2, and NS3 include Christians as well as non-Christians, but its Christological grounds appear to confine NS4 to those who accept that human nature is the creaturely condition for a form of life with God that is given to us in Jesus Christ. This shortcoming would be a serious one if NS4 were entirely esoteric. As it is, however, it is readily intelligible to those who are familiar with NS2 and NS3, and its differences from the latter can be explained in two ways that Christian ethics, relying on God's revelation in Jesus Christ, may follow in accounting for its differences from forms of ethics that ignore that revelation or place less emphasis on it. This chapter began with the more familiar of the two ways, introducing NS4 as a conception of the normative status of human nature that avoids certain shortcomings of NS2 and NS3. But it is also possible, from the standpoint of NS4, to regard NS2 and NS3 as partial approximations to NS4, each accounting for aspects of it yet falling short due to their partiality and their failure to grasp the Christological ground which NS4 articulates, yet each also providing instruction and correction to versions of NS4 that, as we have now seen, are themselves incomplete and subject to their own distortions.

Conclusion

Human biological nature is a part of the world created by God, and to refer to something as created by God is, at least in Christian ethics, to imply that it is not normatively indifferent. The prospect of biotechnological enhancement in its many forms lends urgency to the question of the normative status of human biological nature, which is now susceptible to intentional action in ways and to degrees that are unprecedented. This book has now examined at considerable length four versions of the claim that normative status attaches to human nature in the context of biotechnology, and the reader who has made it to this point is surely entitled to a report on the results of the examination. Each of the four preceding chapters ended with a summary and evaluation of its respective version of this claim, including what I take to be the contributions and shortcomings of that particular version. This conclusion will presuppose those particular summaries and evaluations and will not repeat them. Instead it will convey what I think the inquiry allows us to conclude about the normative status of human nature in relation to biotechnology, at least as Christian ethics is concerned with it, and why I think a focus on the normative status of human nature is compatible with the bioethical principles that are most frequently invoked in the ethical evaluation of biotechnological enhancement.

It was no doubt clear by the end of Chapter 5 that my strongest sympathies lie with NS4. That is no surprise as NS4 is my own construction, and I introduced it into the debate over the normative status of human nature because I found the existing positions unsatisfactory in one way or another. However, it was also clear by the end of Chapter 5 that I do not regard NS4 as "the" answer to the question of the normative status of human nature in the context of biotechnology. For one thing, NS4 as I presented it appropriates indispensable insights from its rivals, and its own shortcomings are subject to correction by them. Moreover, there are flaws in the two versions of NS4 I examined. But most fundamentally, the normative significance of human nature is highly complex, and I doubt that any one position can do full justice to it, whether in the context of biotechnology or in any other context. The legitimate contributions of each version of the claim that normative status attaches to human nature ensure that each of them has a legitimate place in the evaluation of biotechnological enhancement in Christian ethics. It is therefore incumbent on me to show how I understand their contributions in relation to one another and why I am persuaded of the relative superiority of NS4. These tasks are best met by pairing NS1 with NS3 and NS2 with NS4, as I will now do.

NS1 emphasizes that creation is both good and a finished work, and although its status as finished pertains to its order rather than to the things that are ordered by it (which things come into existence, go out of existence, and undergo change), its rejection of the determination of human biological characteristics by intentional human action (such as the selection or design of the genetic characteristics of children by their parents) lends support to the notion that human nature does not need to become something other than it now is for its creaturely nature to be complete. NS1 also insists that the eschatological transformation that God destines for human nature does not run along the lines marked out by biotechnological progress or any other historical process. Whatever the eschatological transformation that awaits our nature will involve, it will not be a continuation of the alteration of cognitive, perceptual, emotional, or physical functions or traits by genetics, psychopharmacology, neurology, organic-digital interfaces, or any form of biotechnology. NS1 conflicts with NS3 on both points, and in my view NS1 is right on both.

However, it is also true that the status of creation as a finished work must accommodate the variation and change that characterize human nature as the product of biological processes and unintentional human activity, as NS1 and NS3 both do. As I have noted many times throughout this study, our nature from the beginning until now has been the product of these biological and cultural factors. It follows that intentional action to alter human biological characteristics cannot be ruled out on the grounds that simply by changing human biological characteristics, it violates the status of creation as a finished work. But, NS3 goes beyond merely accommodating variation and change (which NS1 also does) with its claim that variability (indeterminacy) and change (open-endedness) are normatively significant features of human nature. With this claim, NS3 serves a vital moral task by effectively undermining the racism, sexism, and ableism that thrive on constructions of human nature as fixed, unchanging, and

intact. Indeed, by finding normative significance in features that undermine these "-isms" it is potentially more effective against them than are
the many theories that eschew normative discourse when speaking of
human nature under the questionable assumption that problematic normative constructions can be dismantled by simply deconstructing human
nature or describing it more adequately. Moreover, NS3's claim that the
susceptibility of human nature to alteration (malleability) is normatively
significant suggests that (to use the language of NS1) in bringing the things
that are ordered to the state God has purposed for them, God may work
in part through intentional human action and not only (as NS1 holds)
through biological processes and unintended human actions. It thereby
avoids the problem that NS1 faces in having to explain why God is free to
work through everything except intentional human action, excluding only
the latter from participation in God's work. On both points, NS3 proves
itself superior to NS1. At the same time, however, we saw that NS3 finds
it difficult to show how the status of creation as good and finished applies
to our nature as it is, with its distinctive characteristics (and not merely its
indeterminacy, open-endedness, and malleability, by virtue of which it can
become something other than it is). If we cannot see or experience creation
as in some way a good and finished work in the distinctive characteristics
of our nature as it is, its normative significance is imperiled. That NS3 has
difficulty with this affirmation counts against it.

Thus far, it appears that NS1 and NS3 mutually correct one another's
shortcomings and that all that remains for Christian ethics to do in this
domain is to formulate a position that reflects their mutual corrections.
However, NS1 and NS3 both fall short at a fundamental level insofar
as they rely too heavily on their respective versions of the metanarrative
of creation and eschatology to do their normative work and neglect to
inquire adequately into the good that God purposes for humans in bringing a creature of their nature into being and destining it for eschatological
fulfillment. NS1 falls short in this way by supposing that the question of
the good to be achieved through biotechnological intervention is simply
not relevant to the evaluation of the determination of biological characteristics. If human nature is off-limits to intentional action to determine
it, then the question of the good that is to be achieved by intentional
action appears to be a moot one. This result is salutary to the extent that it
expresses the principle that the biological nature of others is not at our disposal. We saw, however, that the very grounds on which NS1 attempts to
rule out biotechnological determination (namely, the orderings of parents
and children to one another and respect for the child as a person or as the

recipient of unconditional love) were unable to rule out the possibility that determination of a child's characteristics may fulfill a broad positive duty of parents to benefit their children – a duty that can be derived from the teleological ordering of parents to the good of their children (O'Donovan) or from their transforming love for their children (Sandel), and one that is at least not inconsistent with respect for the future consent of the child or with her ultimate equality with her parents (Habermas). In other words, the very principles by which subscribers to NS1 seek to rule out intentional determination of the biological characteristics of children by their parents seem to allow a limited role for such determinations to the extent that they require (or at least permit) parents to benefit their children (and assuming that the interventions do indeed benefit them). Meanwhile, in the case of NS3, if indeterminacy, open-endedness, and malleability are taken to be the only normatively significant features of human nature, then human nature is made available for biotechnological enhancement without any basis for determining what counts as a true improvement. The result, as we saw, is that the question of the good is defaulted to biotechnological progress (Hefner) or to all-purpose goods that instantiate the contemporary liberal notion of the good as that which a person chooses for herself (Peterson) – or it is answered by appealing to something beyond nature (Tanner), which is what NS4 does.

In contrast to NS1 and NS3, NS2 and NS4 place their normative weight not on the creation-eschatology metanarrative but on a notion of the good that God has purposed for human nature in creating it and destining it for eschatological fulfillment. There is a good that God prepares us for in endowing our nature with its distinctive characteristics and capacities and destining it for eschatological fulfillment, and the creation-eschatology metanarrative is the story of how God realizes that good. In my view, if NS2 and NS4 are superior to NS1 and NS3, it is on this ground. However, NS2 and NS4 differ over the relation of that good to human nature. Is the good that God wills for humans in creating them intelligible in terms of their creaturely nature alone or only in reference to something that determines its meaning or purpose from outside it? NS2 rightly identifies the goodness of creation with the suitability of our nature as created by God to our good, but we saw that the claim that our good is intelligible in terms of our nature alone poses a problem for NS2 when we consider the prospect of biotechnological alteration. Why should we be satisfied with the goods of our nature as it is if we can alter our nature in accordance with goods that are, thanks to the malleability of human nature and the progress of biotechnology, now live options for us? The suitability of our nature to our good and the intelligibility of the good as the fulfillment of our nature now become questionable, and with this circumstance NS2 faces a crisis unless (abdicating to NS3) we identify the suitability of our nature to its good with its capacity to become something other than it now is and accept a dialectical relationship between extrapolated goods and alterations of our nature, or (following Kass and Nussbaum) demonstrate that goods that require us to alter our nature are inferior to those that fulfill our nature in its present state.

We saw, however, that Christian ethics has good reasons for denying that human nature is a sufficient ground for the human good and points us instead to a ground that transcends our nature, bringing us to NS4. But to say that the ground of our good transcends our nature is not to say that it is realized apart from our nature and thus requires us to abandon our nature or change it into something different. The good that transcends our nature, namely, life with God, is fulfilled in Jesus Christ, who shares our nature, and we enjoy it in him by the grace of God working in or through our nature. It is *as* creatures with this nature that God willed for us to enjoy life with God, which is to be realized *by* the work of grace, and whatever our eschatological fulfillment consists in, it will be the fulfillment of this creature with this nature. In this way, NS4 explains how our nature in its present state equips us for the good that God has purposed for us in creating us, and destining us for eschatological fulfillment, thus accounting for both the goodness of creation and its status as a finished work while also leaving room for a final fulfillment that is continuous with creation as a good and finished work even as it goes beyond it. With its transcendent perspective, NS4 thus corrects the shortcoming of NS2 while preserving in its way key claims of NS2 regarding the adequacy of our nature to our good.

NS2 (especially as represented by Kass and Nussbaum) and NS4 jointly contribute two crucial insights regarding biotechnological enhancement that are worth stressing as I close this review. The first insight is that the suitability of our nature to our good requires only the presence of the relevant characteristics and capacities (characteristics such as mortality for Kass and Nussbaum or sociality, body-soul composition, and temporality for Barth; and capacities such as reason, will, and sensibility for Tanner). Biotechnological enhancement is not capable of contributing anything to the suitability of our nature to our good by effecting qualitative or quantitative changes to our characteristics and capacities. However, this same conviction implies that the suitability of our characteristics and capacities to our good is compatible in principle with a wide range of variability in their particular instantiations in individuals and in human nature over time. Just as their alteration – whether it is by intentional actions or as the unintended effect of other actions – does not render them more suitable to our good, it does not render them less suitable to it either. If it is the mere presence of these characteristics and capacities that equips our nature for life with God, and if their manifestations in individual humans are widely varied and changing, then the question for biotechnological enhancement is not whether it renders our nature more or less suitable to the life with God for which God created it. Rather, the question is whether it actualizes possibilities of our nature that instantiate this meaning and purpose. Can a proposed biotechnological enhancement, such as life extension, be understood and affirmed, not as rendering our nature more suitable to our good, but as actualizing possibilities of our already-suitable nature that instantiate our good? This is the question NS4 poses not only to particular biotechnological enhancements but also to any activity by which we act on our natural characteristics and capacities. Given that our existing characteristics and capacities are adequate to their meaning and purpose, which possibilities of these characteristics and capacities should we actualize as instantiations of the meaning and purpose of our nature (and thus perhaps also as anticipatory signs of, though not steps toward, its eschatological fulfillment)?

The second insight is that what biotechnology might do to our natural characteristics and capacities is not the only matter that is at stake. As we saw in the positions of Kass and Nussbaum on mortality and emphasized in drawing the implications of Barth's position for life-extension technology, the goods of human nature are goods of action and are not simply enjoyed automatically by the possession of the relevant characteristics and capacities. In the case at hand, humans do not enjoy or experience the meanings and purposes that are related to the bounded life span simply by being (and remaining) mortal. Just as those meanings and purposes might be enjoyed and achieved by people who live vastly longer lives than we now do, they are also forfeited by people who, even without adding a day to their lives, inculcate in their attitudes and express in their actions the meanings and purposes of endless duration of life. The meaning and purpose of our nature are enjoyed or experienced only as they are inculcated in our attitudes and expressed in our actions, so that in our ethical evaluations of biotechnological enhancement we must ask of any biotechnological intervention into our characteristics and capacities not only whether the possibility it actualizes (for example, an extended life span) genuinely instantiates the meaning and purpose of our nature as God's creation but also whether the action that brings the possibility about inculcates and expresses this meaning and purpose. The human creaturely good is a good of human action. If it is true that our nature makes its meaning or purpose available to us simply by virtue of the incidence in our nature of the characteristics and capacities we now have, it is also true that we actualize that meaning or purpose only as we express it in our actions, making it genuinely our own as we live by it. This insight of Kass and Nussbaum is crucial: It is not just what we do to our nature that matters but also how we actualize the meaning and purpose of our nature in our attitudes, actions, and practices.

I now turn to the compatibility of claims about the normative status of human nature with other factors in the ethical evaluation of biotechnological enhancement. Far from excluding or marginalizing other considerations in the ethical evaluation of biotechnological enhancement, the focus on the normative status of human nature recognizes the normative force of the standard bioethical principles of autonomy, safety, and fairness. Recognition in this case may go well beyond mere acknowledgment of these principles as external constraints. As I will briefly demonstrate, these principles may be derived from a conception of the normative status of human nature in a way that avoids a common (and in my view legitimate) line of criticism that is often directed at them. What I have to offer here is no more than the most minimal sketch of a position; it is in no way adequate to the issue. A thorough treatment of the relationship of the normative status of human nature and principles such as autonomy, safety, and fairness in the evaluation of biotechnological enhancement technologies is beyond the scope of this book, which as I have noted from the outset is not a general inquiry into the ethics of biotechnological enhancement but a particular inquiry into an important yet widely misunderstood aspect of it. My aim here is therefore simply to demonstrate that my inquiry into the normative status of human nature is not in conflict with the major principles of bioethical inquiry and is not proposed as an alternative to these principles but is compatible with them and indeed grounds them in a way that avoids certain problems that otherwise attend them.

To begin, it is worth stressing that most bioethicists, whether they are in the mainstream or in the countercurrent regarding biotechnological enhancement, agree that even in their current state of development, human enhancement technologies raise important ethical questions related to autonomy, safety, and fairness. In most bioethical debates, appeals to autonomy, safety, and fairness aim at avoiding certain wrongs and harms that might accompany the development and use of these technologies. Will these technologies involve coercion? If they are found to confer competitive or social advantages, people may be (or may feel) pressured to use them. Will they involve inordinate risks? As with any new technology, not much is known about the long-term effects of most of them. Will their distribution be unfair, and will it exacerbate existing unfairness? Almost certainly at first, and possibly in perpetuity, there will be cost barriers as well as social barriers to access as well as diversions of resources from more urgent human needs. To deny or even underestimate the importance of these questions to the ethical evaluation of biotechnological enhancement would be morally irresponsible.

Some critics, however, consider these questions to be superficial, arguing that in their focus on avoiding these wrongs or harms they merely smooth the path of techno-capitalism and the ethos of expressive individualism and economic productivity it promotes. In my view, whenever considerations of autonomy, safety, and fairness are taken to be sufficient for the ethical evaluation of biotechnological enhancement, as they often are in mainstream bioethics, this assessment is on the mark. In such cases it is plausible to suspect that these principles facilitate the role of biotechnology in rendering human nature a site of expressive individualism and economic productivity. As such, autonomy is reduced to the principle that each person lives a life of her own choosing (which is reflected in the claim that the selection and alteration of biological characteristics should promote all-purpose goods that facilitate any way of life a person would eventually choose); safety is reduced to a principle of risk management; and fairness is reduced to a principle of just competition. The suspicion that in these reduced-to forms these three principles serve the late-capitalist ethos of expressive individualism and economic productivity is one that I fully share. In my view, any adequate ethical analysis of biotechnological enhancement would assign a major role to a critique that runs along precisely these lines. However, it is possible to recognize in autonomy, safety, and fairness partial, and partially distorted, specifications or approximations of moral values that protect vital aspects of human nature as Christian ethics understands the latter, and on that ground to affirm them as necessary to ethical evaluations of technology without taking them as sufficient. Thus, autonomy may be affirmed as a specification of respect for the distinctively human form of agency, by virtue of which characteristically human goods are not attainable by instinct but must be reflectively chosen; safety as a specification of respect for the bodily nature of humans, which renders them vulnerable to injury and death in their pursuit of their good; and fairness as a specification of respect for the interdependence of humans, who attain their good in cooperation with one another. These specifications are, to be sure, highly general, but in case the reader suspects that they are merely ad hoc contrivances, it will be helpful if I point out that I have formulated them with Barth's threefold conception of human nature in mind. Barth's notion of temporality, with its emphasis on the singularity of responsibility and moral vocation, supports and indeed requires a space for agency that is marked out by respect for autonomy; his notion of body-soul composition requires respect for bodily integrity that is partially secured by the principle of safety; and his notion of relationality, with its insistence that no one realizes humanity in himself or herself as a self-enclosed subject but only face to face with one's fellow humans, implies interdependence in which no one's good is pursued apart from the good of the other. (Of course, other conceptions of the normative status of human nature are also capable of grounding these principles with roughly the same content as this Barthian conception endows them. If I prefer this Barthian grounding, it is because I regard it as exemplary, not because I regard it as exclusive.)

A thorough treatment of the ethics of biotechnological enhancement would require extensive elaboration and defense of these specifications of these principles along with a critique of their conscription in the ethos of expressive individualism and economic productivity. However, it should be clear how, at least in Barth's case, the aspects of human nature by virtue of which normative status attaches to it not only support but indeed ground the most commonly invoked bioethical principles. Critics will point out that much is either lost or distorted in the translation from the Christian to the liberal idiom, and they will be right. However, that is a reason for Christian bioethicists to redouble their efforts to recover what has been lost and correct what has been distorted in this translation, not to abandon their principles on the grounds that others have mishandled them.

I conclude with the central question of this book: In an age of biotechnological enhancement, how do we actualize the meaning and purpose of the biological nature God has given us in bringing us into existence as the creatures we are? To keep our nature off-limits (NS1) or to default to its susceptibility to intervention (NS3) are inadequate responses. We have no choice but to determine whether biotechnological interventions actualize or subvert the meaning and purpose of our creaturely nature (NS2, NS4). In a morally well-ordered society, principles of autonomy, safety, and fairness will establish appropriate ethical conditions for Christians to make these determinations. But to make these determinations well will require a great deal of discernment, deliberation, and prayer. My hope is that this book has contributed something to that task.

Appendix

Claims that normative status attaches to human nature typically focus on some feature(s) of human nature (for example, its immunity against intentional human determination, as in NS1, or its indeterminacy, open-endedness, and malleability, as in NS3) or some role that it plays (for example, as the ground of human goods or rights, as in NS2, or as the condition of life with God, as in NS4) as normatively significant. However, there is one context in which it is sometimes claimed that normative status attaches to human nature not with respect to any feature(s) of it or any role it plays, or even by virtue of any particular value or purpose that it has, but simply as such. That context has to do with the prospect that biotechnological alteration of human nature might develop to the point that those who have undergone it are no longer human but have become something else. I will refer to this as the prospect of the "posthuman," by which I mean the succession of human nature, whether in the case of individuals or populations, by another kind of nature that is brought about by the application of technology to humans. (This prospect is distinct from that of direct production of nonhuman entities by technology, which raises its own questions.) In light of this prospect, one might claim that human nature has value simply as human, so that normative status attaches to human nature simply as such. Based on this claim, one might go on to argue that we have reason, and are perhaps even obligated, to refrain from biotechnological alterations that change (or could change) human nature into something else, even in the face of reasons that favor those alterations.

The prospect of the posthuman, so understood, raises many questions. What condition of human nature would count as a posthuman one? How would we recognize it? What technologies might plausibly bring it about? What moral status would a posthuman being have? These questions are all important, but from the perspective of this book the crucial question (and the only one I will address) is whether the prospect of the posthuman is ruled out by the normative status that attaches to human nature or is instead permitted or even welcomed by it. To put it differently, do the claims regarding the normative status of human nature that this book has considered entail the claim that normative status attaches to human nature as such, so that (other things equal) the prospect of the posthuman is ruled out, or do they oppose that claim?

Before turning directly to this question, it is worth pausing to consider whether it is a question that is worth asking. For one response to the prospect of the posthuman as I have defined it is to deny that it could ever come about. This denial is expressed in two opposite ways, both of which we have already encountered. One way is to define "human nature" so broadly that virtually no degree or kind of change to it would amount to a transformation of it into something else. If no biotechnological changes to human nature would count as going beyond human nature, then the prospect of the posthuman is a red herring. This denial has some initial plausibility if one holds to a common interpretation of the classical definition of man as a rational animal. If "rational animal" and "man" are coextensive, then it seems that little if anything that biotechnology would plausibly bring about will change humans into something else, as it is unlikely that biotechnological changes to humans will produce something that is not a rational animal. Of course, the so-called upload scenarios proposed by some cybernetic futurists, in which one transfers information from one's brain to a computer, could (if successful) result in a rational nonanimal, and that would seem to count as posthuman. However, assuming that the uploader retains the original, so to speak, in her brain, these scenarios are best understood as ones in which humans make nonorganic copies of themselves (or of certain aspects of themselves), not as transformations of their nature into something else. In any case, one might argue against this first way of denial by appealing to the existence of different kinds of rational animals. Examples might be other members of the genus *Homo* and perhaps other hominins or even members of the genus *Delphinus* (that is, dolphins). However, this argument may not convince defenders of this first way, who might deny that other rational members of the genus *Homo* belong to a different kind from ours and might protest that dolphins, though intelligent, are not rational. A better argument would appeal to the possibility of intelligent living things elsewhere in the universe. If the existence of rational animals whose rational capacities and animal nature vastly differ from ours, and who are not connected with us through any evolutionary process, is logically and empirically possible, then there may be very different kinds of rational animals, and it is therefore at least conceivable that biotechnologically induced changes could amount to a transformation of humans into beings of a different kind even as the transformed beings are, like humans, rational animals. Returning to the terrestrial context, Chapter 3 argued that if human biotechnological alteration eventually reaches a point where the enhanced and unenhanced exhibit significantly different behaviors and characteristics and sort into separate social groups, the best account of these traits and tendencies might well explain them as possibilities of distinct kinds of beings. These arguments make clear that it begs the question to rule out the prospect of the posthuman *a priori*, on the grounds that any biotechnologically altered beings, however different their characteristics, behaviors, and forms of life, would count as members of the same kind as the human beings from which they emerged as long as they are still rational animals. In sum, the prospect of the posthuman cannot be denied by simply defining the human kind broadly enough to accommodate all the changes to human nature that biotechnology would plausibly bring about.

The other way to deny that the prospect of the posthuman as I understand it could ever come about is to deny that there is any human nature that could be transformed into something else. Those who exemplify this second way of denial argue that due to the absence of clear or fixed boundaries between humans and nonhumans or to the coevolution of the human organism and technology, we already are posthuman and perhaps always have been – meaning not that our human nature has already transformed into something else (which is what *posthuman* means on my definition) but that we now are and perhaps always have been without a human nature. If there is no such thing as human nature, then nothing that biotechnology does to us can change human nature into something else. However, as I argued in Chapter 1, the claims that there are no clear or fixed boundaries between humans and nonhumans and that humanity and technology have coevolved (both of which are highly plausible claims that I endorse) do not entail the claim that there is no such thing as human nature. To suppose that the former claims do entail the latter one is to make the questionable assumption that the nature of something must be fixed and unchanging – an assumption that is routinely made by those who understand "posthuman" in this sense. If we drop that problematic assumption, then the most plausible view is that blurred boundaries with nonhumans and coconstitution by technology are characteristics of human nature (though of course they need not be exclusive to human nature). To be sure, these characteristics of human nature render it difficult to change the latter into something else (which would of course be rather easy, at least in theory, if

all one needed to do was change one unique and essential trait) and difficult to determine when any such change has occurred. But in principle,
biotechnology could on this view change human nature into something
else. It follows that the prospect of the posthuman as I have defined it cannot be ruled out on the grounds that there is no human nature that can be
transformed into something else.

Now that its relevance has been confirmed, let us return to our question: Do the claims regarding the normative status of human nature that
this book has considered entail the claim that normative status attaches to
human nature as such, so that the prospect of the posthuman is ruled out,
or do they oppose that claim? The claim that normative status attaches to
human nature as such in some way that renders illicit technologies that
would transform it into something else has been explicitly rejected on at
least three distinct grounds, which I refer to as "transhumanist," "welfarist," and "perfectionist." In all three cases the prospect of the posthuman
is understood roughly as I understand it, namely, as a state in which technology has transformed human nature (whether in the case of individuals
only or at the population level) into something else, and the claim that this
prospect violates some normative status that allegedly attaches to human
nature as such is rejected. I will first consider these three rejections, then
turn to the results of our examination of the four versions of the claim that
normative status attaches to human nature.

Transhumanists explicitly promote the prospect of the posthuman.
Humanity+, which is the most prominent transhumanist organization,
defines *transhumanism* as "a way of thinking about the future that is based
on the premise that the human species in its current form does not represent
the end of our development but rather a comparatively early phase" and
accordingly envisions "future beings whose basic capacities so radically
exceed those of present humans as to be no longer unambiguously human
by our current standards."[1] The wording carefully avoids commitment to
one or the other answer to the question whether these beings would in fact
still be human. Hesitancy on this matter is appropriate, given the absence
of clear boundaries between humans and nonhumans and the consequent
difficulty (and perhaps even genuine ambiguity) involved in distinguish-
ing one from the other. But in either case, the posthuman state toward
which transhumanists look lies in the future and is not merely a description of our current relationship to technology. "Radical technological

---

[1] Humanity+, "Transhumanist FAQ," http://humanityplus.org/philosophy/transhumanist-faq/
(accessed August 9, 2016).

modifications to our brains and bodies are needed" if our condition is to count as posthuman in the sense meant by transhumanists. What distinguishes transhumanists from the welfarists and perfectionists whom I will go on to describe is the value they ascribe to future states in which, as they anticipate, goods that are not yet knowable or even imaginable will be accessible to us or our descendants. As transhumanists understand it, the good is, in effect, the as-yet-unknown correlate of the radically enhanced capacities that can recognize it as good and enjoy it. Transhumanists, then, are committed to keep pushing biotechnological enhancement forward with the promise that there will be great goods to enjoy, whether for ourselves or (more likely) for our successors.

Welfarists do not argue (as transhumanists do) that we should try to bring about beings who can experience superior goods. Rather, they accept the prospect of the posthuman as the possible concomitant of the promotion of the well-being of existing humans. As we saw in Chapter 3, bioethicists such as Allen Buchanan, Jonathan Glover, and John Harris argue that we should pursue biotechnological enhancement if it genuinely benefits people and that whether their nature changes into something else as the accumulated result of beneficial enhancements is of no moral consequence. For these and other welfarists there are of course many qualities of human nature that we should value and preserve from loss. But these qualities are valuable insofar as they are beneficial to those whose qualities they are, not because they are part of their nature. If it becomes possible to enhance these qualities so that they become even more beneficial or to alter or eliminate other qualities that are not beneficial, then we have in principle an obligation to do so, and whether those who undergo these changes become something other than human as a result is not morally significant.

Finally, perfectionists such as Philip Hefner and James Peterson, whom we met in Chapter 4, agree with transhumanists that our nature in its present form is not the final stage of our development, but for them the transformation of human nature occurs on the way to the teleological or eschatological destiny of creation as a divine project in which creatures cooperate; it is not merely the occasion for superior experiences and enjoyments. In principle, we can distinguish between perfectionists as the inheritors of a classical teleological position, yet without the fixed natures that position assumed, from transhumanists as the inheritors of a modern liberal position, yet without the natural limitations that position places on our desires or preferences. In practice, however, this distinction is a blurry one, and there is considerable overlap between the two. In any case, to attach normative status to human nature as such is for perfectionists to block the teleological or eschatological destiny of creation rather than to cooperate with God in moving it toward its realization.

The claims of NS1 and NS2 that have survived my critical examination leave us with few if any decisive reasons to oppose transhumanism, welfarism, and perfectionism. In the case of NS1 the opposition would have to be grounded in the wrongness of altering the generic orders that comprise creation as a finished work (notwithstanding the changes to the things that are ordered) or in the wrongness of the transformation of human nature into something else as an instantiation of a stance toward nature as mere matter at the disposal of the human will. But if things that are ordered can come into and go out of existence without changing the orders (so that the arrival of the posthuman at a certain point in time is no less consistent with its holding a place in an unchanging generic order than was the arrival of the human at its point in time), and if transformations of human nature into something else are the effect of actions that genuinely benefit those who undergo the transformations in accordance with generic relations that require or at least permit mutual benefit among the beings that are generically related to one another (so that the posthuman results from acts that accord with the ordering of creation into generic kinds and not from mere imposition of form on matter), then it is unclear that either of these wrongs will have been committed if the prospect of the posthuman materializes. Meanwhile, I argued in Chapter 3 that unless it adopts an explicit eschatological perspective, NS2 has difficulty ruling out either the welfarist scenario of a gradual accumulation of beneficial changes that result in a different kind of creature or the transhumanist desire to experience yet unknown goods. Thus far, then, it appears that there is nothing to stop transhumanism, welfarism, and perfectionism from denying that normative status attaches to human nature as such in a way that rules out the prospect of the posthuman.

However, my critical examinations of NS3 and NS4 do in fact leave us with grounds for extending normative status to human nature as such. I argued in criticism of NS3 that the goodness of creation and its status as a finished work are compromised if they are found only in the capacity of our nature to become something other than it now is and not in our nature in its present state. I also showed how for NS4 the characteristics and capacities that constitute the present state of human nature are those in or through which life with God is enjoyed. If enjoyment of a particular form of life with God is the reason for God's creation of humans in the first place, and if the characteristics and capacities that comprise human nature in its present state are those in or through which life with God is enjoyed (in which cases human nature in its present state, taking into account, of course, its distortion due to sin and the fulfillment that is its eschatological destiny, instantiates the goodness of creation and its status as a finished work), then there is a reason, at least for Christian ethics, for humans to retain the characteristics and capacities they now have rather than pursuing biotechnological alterations that might bring about different characteristics and capacities. On this view, it is *these* characteristics and capacities, and not any other ones, that grace works on to bring us into full enjoyment of life with God, even if the work of grace generates possibilities that are not inherent in these characteristics and capacities as created by God (that is, if their eschatological fullness is greater than their original creation). It is in this way that Christian ethics can affirm both that our created nature in its present state instantiates the good and finished character of creation (insofar as there is no need for new or different characteristics and capacities) and that the eschatological fullness of life with God involves a transformation of our created nature that takes it beyond its created state (insofar as grace works on or with those characteristics and capacities to generate possibilities that are not inherent in their created state).[2]

I close with three observations that follow from the general point I have just made. First, biotechnological enhancement of our present characteristics and capacities is not ruled out by these considerations. Although there is no sharp line between them, there is in principle a distinction between changes *to* the characteristics and capacities that comprise human nature and changes *of* those characteristics and capacities to something else. It was for sound reasons that Chapter 5 left open the possibility that biotechnological enhancement, while not rendering our nature any more suitable to the life with God for which God created us, might actualize possibilities of our nature that are in accordance with that meaning and purpose of our nature by changes to certain characteristics and capacities. And even if that possibility never materializes, it remains true that changes to human characteristics and capacities have been occurring for as long as there have been human beings.

Second, our life with God is constituted in Jesus Christ (it is indeed participation in his own life with God), and it is by virtue of Christ's assumption of our human nature that we live our life with God. This observation

---

[2] From this perspective, transhumanists are justified in their valuation of goods that we are as yet incapable of enjoying or even fully imagining but mistaken in their assumption that we must acquire dramatically different characteristics and capacities to enjoy these goods. Rather, it is our participation in Christ and the work of grace that actualizes it that render us capable of enjoying them.

suggests that posthumans might not share in the life with God that Christ in his humanity realizes and makes accessible to those who share the nature he has assumed. After all, it is in and through the characteristics and capacities that other human beings share with Christ, and not some others, that they enjoy the life with God which they have in Christ. However, this suggestion must be qualified in two ways if we are to avoid misunderstanding it. First, it does not follow that posthumans who have different characteristics and capacities from ours would be excluded from life with God. While Christ assumed humanity in a special and unique way, all flesh, and indeed the whole of creation (according to Colossians 1:15–20, "all things, whether on earth or in heaven"), participates in his incarnation and therefore, in some way, in his life with God. More generally, all creatures have been created to commune with God in ways that are appropriate to their nature, and the same would be true of posthuman creatures. Second, it is at least plausible to suppose that posthumans would have some share in the distinctively human form of life with God by virtue of their descent from humans. We should not rule out the possibility that something of the determination of the progenitors for this form of life with God would be inherited by their progeny. Despite these qualifications, however, Christian theology cannot avoid the implications of Christ's becoming incarnate in human form. The life with God that is lived in and through human characteristics and capacities is rendered special and unique by being the life God chose to live, as well as the life through which other creatures come into the fullness of their life with God (Rom. 8:22; Heb. 2:7–9). To be human is to participate in this special and unique vocation in creation and redemption, and to become something other than human would be to forfeit full participation in this role, whatever might be retained of it and whatever might be gained by the superiority of the new characteristics and capacities to those of humans.[3]

Finally, as the last remark implies, to assert that the highest form of life with God is the one that is lived in and through the characteristics and capacities of human nature in its present form – an assertion that, if it is indeed warranted, is warranted on the grounds that it is the life with God that is realized in Christ – is not to assert that human characteristics and capacities are superior to those of other creatures, including posthumans, or that other creatures, including posthumans, are not worthy of our respect. If they eventually appear, posthumans may well be superior to

---

[3] I am indebted to Andrew Davison for helping me think through the line of thought I have sketched in this paragraph and the issues that pertain to it.

humans in many ways, just as angels also are, yet without being chosen or destined for the life with God for which humans have been created. They should in any case be respected as the kind of creature they are, just as every creature should be respected as the kind of creature it is. There will no doubt be much to admire in them. But we should not be misled by the superiority of their characteristics and capacities. It would simply indicate that the God who did not choose the most outwardly impressive nation for a unique and special role in history had also not chosen the most outwardly impressive species for such a role in the cosmos.

Bibliography

Agar, Nicholas, *Humanity's End: Why We Should Reject Radical Enhancement* (Cambridge, MA: MIT Press, 2010).

Albertson, David, and King, Cabell, eds., *Without Nature? A New Condition for Theology* (New York: Fordham University Press, 2010).

Arendt, Hannah, *The Human Condition* (Chicago: University of Chicago Press, 1958).

Augustine, *City of God against the Pagans*.

On the Literal Interpretation of Genesis.

Baillie, Harold W., and Casey, Timothy K., eds., *Is Human Nature Obsolete? Genetics, Bioengineering, and the Future of the Human Condition* (Cambridge, MA: MIT Press, 2005).

Barth, Karl, *Church Dogmatics*, Vol. III, Part 1 (Edinburgh: T&T Clark, 1958).

Church Dogmatics, Vol. III, Part 2 (Edinburgh: T&T Clark, 1960).

Church Dogmatics, Vol. III, Part 4 (Edinburgh: T&T Clark, 1960).

Church Dogmatics, Vol. IV, Part 3.1 (Edinburgh: T&T Clark, 1961).

Bostrom, Nick, "Transhumanist Values," *Review of Contemporary Philosophy* 4 (2005): 87–101.

Buchanan, Allen, *Beyond Humanity? The Ethics of Biomedical Enhancement* (Oxford: Oxford University Press, 2011).

Buchanan, Allen, Brock, Dan, Daniels, Norman, and Wikler, Dan, *From Chance to Choice: Genetics and Justice* (New York: Cambridge University Press, 2000).

Butler, Judith, *Bodies That Matter* (New York: Routledge, 1993).

Canguilhem, Georges, *The Normal and the Pathological*, Fawcett, Carolyn R., tr., in collaboration with Cohen, Robert S. (New York: Zone Books, 1991).

Centers for Disease Control, *National Vital Statistics Reports* 61.3 (2012).

Coeckelbergh, Mark, *Human Being @ Risk: Enhancement, Technology, and the Evaluation of Vulnerability Transformations* (Dordrecht: Springer, 2013).

Cole-Turner, Ronald, *The New Genesis: Theology and the Genetic Revolution* (Louisville, KY: Westminster/John Knox Press, 1993).

ed., *Transhumanism and Transcendence: Christian Hope in an Age of Technological Enhancement* (Washington, DC: Georgetown University Press, 2011).

Davis, Lennard J., *Enforcing Normalcy: Disability, Deafness, and the Body* (London: Verso, 1995).

"Introduction: Disability, Normality, and Power," in Davis, Lennard J., ed., *The Disability Studies Reader*, 4th ed. (New York: Routledge, 2013), pp. 1–14.

Deane-Drummond, Celia, and Scott, Peter Manley, eds., *Future Perfect? God, Medicine, and Human Identity* (Edinburgh: T&T Clark, 2006).

de Grey, Aubrey, *Strategies for Engineered Negligible Senescence: Why Genuine Control of Aging May Be Foreseeable* (New York: New York Academy of Sciences, 2004).

de Grey, Aubrey, and Rae, Michael, *Ending Aging: The Rejuvenation Breakthroughs That Could Reverse Human Aging in Our Lifetime* (New York: St. Martin's Griffin Press, 2007).

Dehaene, Stanislas, et al., "How Learning to Read Changes the Cortical Networks for Vision and Language," *Science* 330 (2010): 359–64.

Dupuy, Jean-Pierre, *On the Origins of Cognitive Science: The Mechanization of the Mind* (Cambridge, MA: MIT Press, 2009).

Elster, Jon, *Sour Grapes: Studies in the Subversion of Rationality* (Cambridge: Cambridge University Press, 1983).

Foot, Philippa, *Natural Goodness* (Oxford: Oxford University Press, 2001).

Fukuyama, Francis, *Our Posthuman Future: Consequences of the Biotechnology Revolution* (New York: Farrar, Straus and Giroux, 2002).

Glover, Jonathan, *Choosing Children: Genes, Disability, and Design* (New York: Oxford University Press, 2006).

Graham, Elaine, *Representations of the Post/Human: Monsters, Aliens and Others in Popular Culture* (New Brunswick, NJ: Rutgers University Press, 2002).

"Bioethics after Posthumanism: Natural Law, Communicative Action and the Problem of Self-Design," *Ecotheology* 9 (2004): 178–98.

"In Whose Image? Representations of Technology and the 'Ends' of Humanity," in Deane-Drummond and Scott, eds., *Future Perfect?*, pp. 56–69.

Graham, Gordon, "Human Nature and the Human Condition," in Deane-Drummond and Scott, eds., *Future Perfect?*, pp. 33–44.

Grant, George, "Thinking about Technology," in *Technology and Justice* (Concord: House of Asansi Press, 1986), pp. 11–34.

Groll, Daniel, and Lott, Michael, "Is There a Role for 'Human Nature' in Debates about Human Enhancement?," *Philosophy* 90 (2015): 623–51.

Gunton, Colin, *The Christian Faith: An Introduction to Christian Doctrine* (Oxford: Blackwell, 2002).

Habermas, Jürgen, "The Debate on the Ethical Self-Understanding of the Species," in *The Future of Human Nature* (Cambridge: Polity Press, 2003), pp. 16–100.

"Faith and Knowledge," in *The Future of Human Nature*, pp. 101–15.

Haraway, Donna J., *Simians, Cyborgs, and Women: The Reinvention of Nature* (New York: Routledge, 1991).

Modest_Witness@Second_Millenium. FemaleMan_Meets_Oncomouse: Feminism and Technoscience (New York: Routledge, 1997).

The Companion Species Manifesto: Dogs, People, and Significant Otherness (Chicago: Prickly Paradigm Press, 2003).

Harris, John, *Enhancing Evolution: The Ethical Case for Making Better People* (Princeton, NJ: Princeton University Press, 2007).

Hayles, Katherine, *How We Became Posthuman: Virtual Bodies in Cybernetics, Literature, and Informatics* (Chicago: University of Chicago Press, 1999).

Hefner, Philip, *The Human Factor: Evolution, Culture, and Religion* (Minneapolis: Fortress Press, 1993).

Technology and Human Becoming (Minneapolis: Fortress Press, 2003).

Heidegger, Martin, "The Question Concerning Technology," in *The Question Concerning Technology and Other Essays*, translated and with an introduction by William Lovitt (New York: Harper and Row, 1977), pp. 3–35.

Irenaeus of Lyons, *Against Heresies*.

Jeeves, Malcolm, ed., *Rethinking Human Nature: A Multidisciplinary Approach* (Grand Rapids, MI: Eerdmans, 2011).

Jersild, Paul, *The Nature of Our Humanity: A Christian Response to Evolution and Biotechnology* (Minneapolis: Fortress Press, 2009).

Jonas, Hans, *The Phenomenon of Life: Toward a Philosophical Biology* (New York: Harper and Row, 1966).

The Imperative of Responsibility: In Search of an Ethics for a Technological Age (Chicago: University of Chicago Press, 1984).

Kaebnick, Gregory, ed., *The Ideal of Nature: Debates about Biotechnology and the Environment* (Baltimore: The Johns Hopkins University Press, 2011).

"Human Nature without Theory," in Kaebnick, ed., *The Ideal of Nature*, pp. 49–70.

Kamm, Frances, "What Is and Is Not Wrong with Enhancement?," in Savulescu, Julian, and Bostrom, Nick, eds., *Human Enhancement* (Oxford: Oxford University Press, 2009), pp. 91–130.

Kass, Leon, *Toward a More Natural Science: Biology and Human Affairs* (New York: The Free Press, 1985).

"The Wisdom of Repugnance," *The New Republic* 216 (June 2, 1997), pp. 17–26.

Life, Liberty and the Defense of Dignity: The Challenge for Bioethics (San Francisco: Encounter Books, 2002).

"Biotechnology and Our Human Future: Some General Reflections," in Sutton, Sean D., ed., *Biotechnology: Our Future as Human Beings and Citizens* (Albany: State University of New York Press, 2009), pp. 9–29.

Kroes, Peter, and Meiers, Anthonie, eds., *The Empirical Turn in the Philosophy of Technology* (Amsterdam: Elsevier Science, 2001).

Kurzweil, Ray, *The Age of Spiritual Machines: When Computers Exceed Human Intelligence* (New York: Penguin, 2000).

Lewis, C. S., *The Abolition of Man* (London: Macmillan, 1947).

Meilaender, Gilbert, *Should We Live Forever? The Ethical Ambiguities of Aging* (Grand Rapids, MI: Eerdmans, 2013).

Messer, Neil, *Flourishing: Health, Disease, and Bioethics in Theological Perspective* (Grand Rapids, MI: Eerdmans, 2013).

Mill, John Stuart, "On Nature," in Lerner, Max, ed., *Essential Works of John Stuart Mill* (New York: Bantam, 1961), pp. 367–401.

Mittelstrass, Jürgen, "Science and the Search for a New Anthropology," in Jeeves, ed., *Rethinking Human Nature*, pp. 61–69.

Moravec, Hans, *Mind Children: The Future of Robot and Human Intelligence* (Cambridge, MA: Harvard University Press, 1988).

Mueller, Laurence D., Rauser, Cassandra L., and Rose, Michael R., eds., *Does Aging Stop?* (New York: Oxford University Press, 2011).

Newman, Stuart A., "Renatured Biology: Getting Past Postmodernism in the Life Sciences," in Albertson and King, eds., *Without Nature?*, pp. 101–35.

Nussbaum, Martha C., "Transcending Humanity," in *Love's Knowledge: Essays on Philosophy and Literature* (New York: Oxford University Press, 1990), pp. 365–91.

The Therapy of Desire: Theory and Practice in Hellenistic Ethics (Princeton, NJ: Princeton University Press, 1994).

O'Donovan, Oliver, *Begotten or Made?* (Oxford: Clarendon Press, 1984).

Resurrection and Moral Order: An Outline of Evangelical Ethics, 2nd ed. (Grand Rapids, MI: Eerdmans, 1994).

Peters, Ted, *GOD—The World's Future: Systematic Theology for a New Era* (Minneapolis: Fortress Press, 1992).

Playing God? Genetic Determinism and Human Freedom (New York: Routledge, 1997).

Peterson, James C., *Changing Human Nature: Ecology, Ethics, Genes, and God* (Grand Rapids, MI: Eerdmans, 2010).

Porter, Jean, *Nature as Reason: A Thomistic Theory of the Natural Law* (Grand Rapids, MI: Eerdmans, 2005).

President's Council on Bioethics, *Beyond Therapy: Biotechnology and the Pursuit of Happiness* (New York: HarperCollins, 2003).

Rahner, Karl, "The Experiment with Man: Theological Observations on Man's Self-Manipulation," in *Theological Investigations, Vol. 9: Writings of 1965–1967*, Harrison, Graham, tr. (New York: Herder and Herder, 1972), pp. 205–24.

Ridenour, Autumn, "The Coming of Age: Curse or Calling? Toward a Christological Interpretation of Aging as Call in the Theology of Karl Barth and W. H. Vanstone," *Journal of the Society of Christian Ethics* 33 (2013): 151–67.

"Union with Christ for the Aging: A Consideration of Aging and Death in the Theology of St. Augustine and Karl Barth," unpublished PhD dissertation, Boston College (2013).

Sandel, Michael J., "Mastery and Hubris in Judaism: What's Wrong with Playing God?," in *Public Philosophy: Essays on Morality in Politics* (Cambridge, MA: Harvard University Press, 2005), pp. 196–210 (originally published in Malino, Jonathan, ed., *Judaism and Modernity: The Religious Philosophy of David Hartman* [Aldershot: Ashgate, 2004], pp. 121–32).

The Case against Perfection: Ethics in the Age of Genetic Engineering (Cambridge, MA: Harvard University Press, 2007).

Savulescu, Julian, Sandberg, Anders, and Kahane, Guy, "Well-Being and Enhancement," in Savulescu, Julian, ter Meulen, Ruud, and Kahane,

Guy, eds., *Enhancing Human Capacities* (Oxford: Wiley-Blackwell, 2009), pp. 3–18.

Scherz, Paul, "Living Indefinitely and Living Fully: Laudato Si' and the Value of the Present in Christian, Stoic, and Transhumanist Temporalities," *Theological Studies* (forthcoming 2018).

Sharon, Tamar, *Human Nature in an Age of Biotechnology: The Case for Mediated Posthumanism* (Dordrecht: Springer, 2014).

Soloveitchik, Joseph B., *Halakhic Man*, Kaplan, Lawrence, tr. (New York: The Jewish Publication Society, 1983).

Song, Robert, *Human Genetics: Fabricating the Future* (Cleveland: Pilgrim Press, 2002).

"Knowing There Is No God, Still We Should Not Play God? Habermas on the Future of Human Nature," *Ecotheology* 11 (2006): 191–211.

"Technological Immortalization and Original Mortality: Karl Barth on the Celebration of Finitude," in Ziegler, Philip G., ed., *Eternal God, Eternal Life: Theological Investigations into the Concept of Immortality* (Edinburgh: T&T Clark, 2016), pp. 187–209.

Steckel, Richard H., "Health, Nutrition and Physical Well-Being," in Carter, Susan, Gartner, Scott, Haines, Michael, Olmstead, Alan, Sutch, Richard, and Wright, Gavin, eds., *Historical Statistics of the United States: Millennial Edition*, Vol. 2 (New York: Cambridge University Press, 2002), pp. 499–620.

Steinbock, Bonnie, "The Appeal to Nature," in Kaebnick, ed., *The Idea of Nature*, pp. 98–113.

Stiegler, Bernard, *Technics and Time, I: The Fault of Epimetheus* (Stanford, CA: Stanford University Press, 1994).

Tanner, Kathryn, *Christ the Key* (Cambridge: Cambridge University Press, 2010).

"Grace without Nature," in Albertson and King, eds., *Without Nature?*, pp. 363–75.

Teilhard de Chardin, Pierre, *The Phenomenon of Man*, Wall, Bernard, tr. (London: Collins, 1965).

Activation of Energy, Hague, René, tr. (London: Collins, 1970).

Toward the Future, Hague, René, tr. (London: Collins, 1975).

The Future of Man, Denny, Norman, tr. (New York: Doubleday, 2004).

Thweatt-Bates, Jeanine, *Cyborg Selves: A Theological Anthropology of the Posthuman* (Aldershot: Ashgate, 2012).

Verhey, Allen, *Nature and Altering It* (Grand Rapids, MI: Eerdmans, 2010).

Waters, Brent, *From Human to Posthuman: Christian Theology and Technology in a Postmodern World* (Aldershot: Ashgate, 2006).

Christian Moral Theology in the Emerging Technoculture: From Posthuman Back to Human (Aldershot: Ashgate, 2014).

Williams, Bernard, "Must a Concern for the Environment be Centred on Human Beings?," in *Making Sense of Humanity and Other Philosophical Papers* (Cambridge: Cambridge University Press, 1995), pp. 233–40.

Zoloth, Laurie, "The Duty to Heal an Unfinished World: Jewish Tradition and Genetic Research," *Dialog* 40 (2001): 299–300.

"The Ethics of the Eighth Day: Jewish Bioethics and Research on Human Embryonic Stem Cells," in Holland, Suzanne, Lebacqz, Karen, and Zoloth, Laurie, eds., *The Human Embryonic Stem Cell Debate: Science, Ethics, and Public Policy* (Cambridge, MA: MIT Press, 2001), pp. 95–112.

"Go and Tend the Earth: A Jewish View on an Enhanced World," *Journal of Law, Medicine and Ethics* (2008): 10–25.

Index